REVIEW OF THE DRAFT FOURTH NATIONAL CLIMATE ASSESSMENT

Committee to Review the Draft Fourth National Climate Assessment

Board on Atmospheric Sciences and Climate
Division on Earth and Life Studies

Board on Environmental Change and Society
Division of Behavioral and Social Sciences and Education

A Consensus Study Report of
The National Academies of
SCIENCES · ENGINEERING · MEDICINE

THE NATIONAL ACADEMIES PRESS
Washington, DC
www.nap.edu

THE NATIONAL ACADEMIES PRESS 500 Fifth Street, NW Washington, DC 20001

This study was supported by the National Aeronautics and Space Administration under award numbers NNH14CK78B and NNH14CK79D. Any opinions, findings, conclusions, or recommendations expressed in this publication do not necessarily reflect the views of any organization or agency that provided support for the project.

International Standard Book Number-13: 978-0-309-47169-5
International Standard Book Number-10: 0-309-47169-9
Digital Object Identifier: https://doi.org/10.17226/25013

Additional copies of this publication are available for sale from the National Academies Press, 500 Fifth Street, NW, Keck 360, Washington, DC 20001; (800) 624-6242 or (202) 334-3313; http://www.nap.edu.

Printed in the United States of America

Suggested citation: National Academies of Sciences, Engineering, and Medicine. 2018. *Review of the Draft Fourth National Climate Assessment*. Washington, DC: The National Academies Press. doi: https://doi.org/10.17226/25013.

The National Academies of

SCIENCES • ENGINEERING • MEDICINE

The **National Academy of Sciences** was established in 1863 by an Act of Congress, signed by President Lincoln, as a private, nongovernmental institution to advise the nation on issues related to science and technology. Members are elected by their peers for outstanding contributions to research. Dr. Marcia McNutt is president.

The **National Academy of Engineering** was established in 1964 under the charter of the National Academy of Sciences to bring the practices of engineering to advising the nation. Members are elected by their peers for extraordinary contributions to engineering. Dr. C. D. Mote, Jr., is president.

The **National Academy of Medicine** (formerly the Institute of Medicine) was established in 1970 under the charter of the National Academy of Sciences to advise the nation on medical and health issues. Members are elected by their peers for distinguished contributions to medicine and health. Dr. Victor J. Dzau is president.

The three Academies work together as the **National Academies of Sciences, Engineering, and Medicine** to provide independent, objective analysis and advice to the nation and conduct other activities to solve complex problems and inform public policy decisions. The National Academies also encourage education and research, recognize outstanding contributions to knowledge, and increase public understanding in matters of science, engineering, and medicine.

Learn more about the National Academies of Sciences, Engineering, and Medicine at **www.nationalacademies.org.**

COMMITTEE TO REVIEW THE DRAFT FOURTH NATIONAL CLIMATE ASSESSMENT

ROBIN E. BELL (*Chair*), Columbia University, Palisades, NY
BILAL M. AYYUB, University of Maryland, College Park
MICHELLE L. BELL, Yale University, New Haven, CT
DANIEL G. BROWN, University of Washington, Seattle
DANIEL R. CAYAN, Scripps Institution of Oceanography, University of California, San Diego
F. STUART CHAPIN III, University of Alaska, Fairbanks
JOHN B. GATES, The Climate Corporation, Alameda, CA
L. RUBY LEUNG, Pacific Northwest National Laboratory, Richland, WA
JANET PEACE, Center for Climate and Energy Solutions, Arlington, VA
JULIE PULLEN, Stevens Institute of Technology, Hoboken, NJ
LaVERNE E. RAGSTER, University of the Virgin Islands, St. Thomas, USVI
MARY RUCKELSHAUS, Stanford Woods Institute for the Environment, Stanford, CA
SUSANNE TORRIENTE, City of Miami Beach, FL
ELKE U. WEBER, Princeton University, NJ
CATHY WHITLOCK, Montana State University, Bozeman
GARY YOHE, Wesleyan University, Middletown, CT

National Academies of Sciences, Engineering, and Medicine Staff
APRIL MELVIN, Program Officer
ERIN MARKOVICH, Senior Program Assistant/Research Assistant
KIMBERLY DUONG, Christine Mirzayan Science and Technology Policy Fellow

BOARD ON ATMOSPHERIC SCIENCES AND CLIMATE

A.R. RAVISHANKARA (*Chair*), Colorado State University, Fort Collins
SHUYI S. CHEN (*Vice Chair*) University of Washington, Seattle
CECILIA BITZ, University of Washington, Seattle
MARK A. CANE, Columbia University, Palisades, NY
HEIDI CULLEN, Climate Central, Princeton, NJ
ROBERT DUNBAR, Stanford University, Stanford, CA
PAMELA EMCH, Northrop Grumman Aerospace Systems, Redondo Beach, CA
ARLENE FIORE, Columbia University, Palisades, NY
PETER FRUMHOFF, Union of Concerned Scientists, Cambridge, MA
WILLIAM B. GAIL, Global Weather Corporation, Boulder, CO
MARY GLACKIN, The Weather Company, Washington, DC
TERRI S. HOGUE, Colorado School of Mines, Golden, CO
EVERETTE JOSEPH, SUNY University at Albany, NY
RONALD "NICK" KEENER, JR., Duke Energy Corporation, Charlotte, NC
ROBERT KOPP, Rutgers University, Piscataway, NJ
L. RUBY LEUNG, Pacific Northwest National Laboratory, Richland, WA
JONATHAN MARTIN, University of Wisconsin–Madison
JONATHAN OVERPECK, University of Michigan, Ann Arbor
ALLISON STEINER, University of Michigan, Ann Arbor
DAVID W. TITLEY, The Pennsylvania State University, University Park
DUANE WALISER, Jet Propulsion Laboratory, Pasadena, CA

Ocean Studies Board Liaison
DAVID HALPERN, Jet Propulsion Laboratory, Pasadena, CA

National Academies of Sciences, Engineering, and Medicine Staff
AMANDA STAUDT, Director
DAVID ALLEN, Senior Program Officer
LAURIE GELLER, Senior Program Officer
KATHERINE THOMAS, Senior Program Officer
LAUREN EVERETT, Program Officer
APRIL MELVIN, Program Officer
AMANDA PURCELL, Program Officer
YASMIN ROMITTI, Research Associate
RITA GASKINS, Administrative Coordinator
SHELLY FREELAND, Financial Associate
ROB GREENWAY, Program Associate
MICHAEL HUDSON, Senior Program Assistant
ERIN MARKOVICH, Senior Program Assistant/Research Assistant

BOARD ON ENVIRONMENTAL CHANGE AND SOCIETY

RICHARD H. MOSS (*Chair*), Joint Global Change Research Institute, College Park, MD
JOSEPH L. ARVAI, University of Michigan, Ann Arbor
F. STUART CHAPIN III, University of Alaska, Fairbanks
RUTH DEFRIES, Columbia University, New York, NY
HALLIE C. EAKIN, Arizona State University, Tempe
LORI M. HUNTER, University of Colorado Boulder
KATHARINE L. JACOBS, University of Arizona, Tucson
MICHAEL ANTHONEY MENDEZ, Yale University, New Haven, CT
RICHARD G. NEWELL, Resources for the Future, Washington, DC
MARY D. NICHOLS, California Air Resources Board, Sacramento
JONATHAN T. OVERPECK, University of Michigan, Ann Arbor
ASEEM PRAKASH, University of Washington, Seattle
J. TIMMONS ROBERTS, Brown University, Providence, RI
MAXINE L. SAVITZ, Technology/Partnership Honeywell Inc., Los Angeles, CA
MICHAEL P. VANDENBERGH, Vanderbilt University Law School, Nashville, TN
JALONNE L. WHITE-NEWSOME, The Kresge Foundation, Troy, MI
ROBYN S. WILSON, The Ohio State University, Columbus

National Academies of Sciences, Engineering, and Medicine Staff
TOBY WARDEN, Interim Board Director
JENNIFER HEIMBERG, Senior Program Officer
HEATHER KREIDLER, Associate Program Officer
JORDYN WHITE, Program Officer
TINA M. LATIMER, Program Coordinator
LETICIA GARCILAZO GREEN, Senior Program Assistant
MARY GHITELMAN, Senior Program Assistant

Acknowledgments

The Committee thanks Philip B. Duffy, Woods Hole Research Center, and Gerald A. Meehl, National Center for Atmospheric Research, for providing input to the Committee on selected sections of the Frequently Asked Questions Appendix contained in the draft Fourth National Climate Assessment (NCA4) report and Gary Morishima, Quinault Management Center, for providing input on Chapter 15 of the draft NCA4 report on Tribal and Indigenous Communities.

This Consensus Study Report was reviewed in draft form by individuals chosen for their diverse perspectives and technical expertise. The purpose of this independent review is to provide candid and critical comments that will assist the National Academies of Sciences, Engineering, and Medicine in making each published report as sound as possible and to ensure that it meets the institutional standards for quality, objectivity, evidence, and responsiveness to the study charge. The review comments and draft manuscript remain confidential to protect the integrity of the deliberative process.

We thank the following individuals for their review of this report:

Susan Avery, ExxonMobil
Christopher Barker, University of California, Davis
John Doyle, Little Big Horn College; Crow Reservation
Patricia Gober, Arizona State University
Sarah Hobbie, University of Minnesota
Klaus Keller, The Pennsylvania State University
Pankaj Lal, Montclair State University
Arthur Lee, Chevron Corporation
Joseph Majkut, Niskanen Center
Constantine Samaras, Carnegie Mellon University
Perry Samson, University of Michigan
Roger Schmidt (NAE), Syracuse University, IBM Corporation, emeritus
John W. (Jack) Williams, University of Wisconsin–Madison

Although the reviewers listed above provided many constructive comments and suggestions, they were not asked to endorse the conclusions or recommendations of this report nor did they see the final draft before its release. The review of this report was overseen by **Charles F. Kennel,** University of California, San Diego, and **Andrew R. Solow,** Woods Hole Oceanographic Institution. They were responsible for making certain that an independent examination of this report was carried out in accordance with the standards of the National Academies and that all review comments were carefully considered. Responsibility for the final content rests entirely with the authoring committee and the National Academies.

Contents

Summary **1**

Chapter 1. Introduction **5**

Chapter 2. Synthesis of Comments on the Draft Fourth National Climate Assessment **7**

Addressing the Mandate in the Global Change Research Act of 1990 8

Linking Impacts with Response Examples 9

Communicating Report Findings 10

Communicating Key Messages 10

Communicating Uncertainty and Risk 13

Bridging Topics and Scales 14

Highlighting New Developments in Climate Science, Impacts, and Responses 15

Chapter 3. Comments on Each Chapter of the Draft Fourth National Climate Assessment **17**

Front Matter: Report Findings 17

Chapter 1: Overview 19

Chapter 2: Our Changing Climate 25

Chapter 3: Water 28

Chapter 4: Energy Supply, Delivery, and Demand 32

Chapter 5: Land Cover and Land Use Change 34

Chapter 6: Forests 37

Chapter 7: Ecosystems, Ecosystem Services, and Biodiversity 40

Chapter 8: Coastal Effects 46

Chapter 9: Ocean and Marine Resources 49

Chapter 10: Agriculture and Rural Communities 51

Chapter 11: Built Environment, Urban Systems, and Cities 55

Chapter 12: Transportation 58

Chapter 13: Air Quality 61

Chapter 14: Human Health 63

Chapter 15: Tribal and Indigenous Communities 65

Chapter 16: Climate Effects on U.S. International Interests 69
Chapter 17: Sectoral Interdependencies, Multiple Stressors, and Complex Systems 71
Chapter 18: Northeast 75
Chapter 19: Southeast 77
Chapter 20: U.S. Caribbean 79
Chapter 21: Midwest 82
Chapter 22: Northern Great Plains 84
Chapter 23: Southern Great Plains 90
Chapter 24: Northwest 93
Chapter 25: Southwest 96
Chapter 26: Alaska 100
Chapter 27: Hawai'i and Pacific Islands 103
Chapter 28: Near-Term Adaptation Needs and Increased Resiliency 106
Chapter 29: Mitigation: Avoiding and Reducing Long-Term Risks 108
References 113
Appendix A. Comments on Appendix 5: Frequently Asked Questions 123
Appendix B. Line Comments 139
Appendix C. Statement of Task 185
Appendix D. Committee Biographies 187

Summary

Climate change poses many challenges that affect society and the natural world. With these challenges, however, come opportunities to respond. By taking steps to adapt to and mitigate climate change, the risks to society and the impacts of continued climate change can be lessened. The National Climate Assessment, coordinated by the U.S. Global Change Research Program (USGCRP), is a mandated report intended to inform response decisions. These reports are required to be developed every four years and provide the most comprehensive and up-to-date evaluation of climate change impacts available for the United States, making them a unique and important climate change document.

The draft Fourth National Climate Assessment (NCA4) report reviewed here addresses a wide range of topics of high importance to the United States and society more broadly, extending from human health and community well-being to the built environment, to businesses and economies, and to ecosystems and natural resources. The report is being developed by hundreds of experts representing federal, state, and local governments, academia, non-government organizations, and the private sector, with further input from community engagement events and public comment. The scale of this collaboration is rare and impressive and the rich array of perspectives introduced through this process provides an opportunity to develop a foundational climate change report that informs and highlights adaptation and mitigation efforts and serves as a valuable resource for broad audiences.

As part of the NCA4 development process, the National Academies of Sciences, Engineering, and Medicine was tasked with convening a panel of experts to provide an external peer review of the draft report. The Committee to Review the Draft Fourth National Climate Assessment ("The Committee") evaluated the draft NCA4 to determine if it meets the requirements of the federal mandate, whether it provides accurate information grounded in the scientific literature, and whether it effectively communicates climate science, impacts, and responses for general audiences including the public, decision makers, and other stakeholders (see Chapter 1 and Appendix C for the full Statement of Task). The Committee approached this charge by developing overarching feedback on the full draft report (Chapter 2) and providing specific comments for individual chapters (Chapter 3 and Appendix B) and the Frequently Asked Questions appendix (Appendix A of this review report).

The Committee was impressed by the accuracy of information and thorough discussion of the predominant aspects of climate change and impacts presented in the draft NCA4. The 1,506-page draft report provides a strong foundation of climate science and a solid discussion of climate change impacts occurring or likely to occur in the United States. The topics are well-selected and logically organized around key messages. The introduction of new national topic and regional chapters since the Third National Climate Assessment (NCA3) is a welcome addition and improves the comprehensiveness of the assessment. The new national topic chapter, "Sectoral Interdependencies, Multiple Stressors, and Complex Systems" is an excellent addition because it facilitates discussion of the inherent challenges introduced by climate change in interlinked systems. Discussion of this critical topic should also be integrated more broadly across draft report chapters. The expanded discussion provided in regional chapters in the draft NCA4 relative to the NCA3 was found to be one of the draft report's greatest strengths. These regional chapters provide a relatively holistic treatment of relevant climate change impacts and are effective in conveying the complex nature of climate change and the linkage among impacts that extend across sectors and topic areas. Regional chapters where specific examples of adaptation and mitigation responses are given are especially strong. The Committee thinks these chapters will resonate well with readers and draw them more fully into the broader report as they learn about climate change impacts "in their backyard."

The draft report represents an expansive and diverse range of information that will be most accessible to readers when the key messages are conveyed clearly and consistently, when linkages across chapters and topics are provided, and when examples of response actions can be drawn on to support the key messages. By improving the communication of key aspects of the draft NCA4, this document could be further strengthened.

Specific overarching recommendations for improving the draft NCA4 include:

Linking Impacts with Response Examples: Incorporate more examples in the draft NCA4 that highlight new and ongoing adaptation and mitigation activities. These should include actions in the private sector, public-private partnerships, and government at multiple scales.

The Committee found the examples of adaptation and mitigation response actions to be very impactful in the draft NCA4, when they are used. Many new actions have been taken in recent years. Examples of these actions could be provided more widely throughout the draft report to illustrate advancements and provide information on how the impacts of climate change are being addressed.

Communicating Report Findings: Reframe the Overview Chapter of the draft NCA4 to center around the twelve report findings that reflect the impacts and responses that are discussed throughout the draft report.

The Overview Chapter (Chapter 1 in the draft NCA4) is expected to be a go-to for readers who are interested in a short synthesis of the NCA4 contents and should complement the "Report in Brief" that will be developed by the NCA4 authors. The chapter is well written and scientifically accurate, but it places strong emphasis on climate science that is already well covered in Chapter 2 of the draft report. The Overview would be more effective if greater focus were given to the impacts and responses discussed across the draft NCA4 report.

Communicating Key Messages: Key messages should be presented using more explicit and concise language. Examples that align with key messages should be included wherever possible in the supporting text and figures. More of the key messages should be supported by examples of response actions to facilitate solution-oriented communication and information sharing.

The draft NCA4 key messages tend to be long and are sometimes hard to follow. This reduces their impact because readers cannot readily identify the take-home points. Tightening language and further prioritizing which information should be included in the key messages versus the supporting text would improve their effectiveness. Many figures in the draft report are not well connected to the key messages. Because the draft report is shaped around the key messages, figures that are closely aligned with the messages would be most useful. Key messages are also an appropriate place to highlight examples of ongoing and planned response actions.

Communicating Uncertainty and Risk: The various types of risk included in the draft NCA4 should be defined more explicitly. Improved consistency in the types of risk discussed and inclusion across chapters is also recommended.

The draft NCA4 report deals with a broad range of uncertainties and risks inherent to the content of a national climate assessment. While some types of uncertainty and risk are discussed (e.g., likelihood and confidence), improved differentiation and more standardized treatment is needed across the draft report.

Bridging Topics and Scales: Linkages to interrelated topics among chapters should be increased throughout the draft NCA4 to ensure consistent treatment of similar topics and to provide readers with a clearer understanding of how impacts and responses at national to regional scales are connected.

The draft NCA4 covers a wide range of topics that are inherently connected. Because of the structuring of the report into national topic, regional, and response chapters, these connections are often missing from the report, leading to coverage of some topics in only one chapter, or in multiple chapters but in different ways. Improved cross-referencing across chapters would better highlight the interconnected nature of the material, guide readers through the information in a manner that will allow them to explore their topics of interest, and make the report more broadly useful.

Highlighting New Developments in Climate Science, Impacts, and Responses: Authors of the draft NCA4 should explicitly identify significant advancements made since the Third National Climate Assessment, with emphasis on emerging science, impacts, and examples of new response actions.

Since the NCA3 was published, scientific research has continued to advance understanding of climate change impacts and the number of response activities has increased. Distinguishing what is new demonstrates measurable progress that is important for informing the NCA4 audience and may also facilitate more solution-oriented messaging of impacts and responses across the report.

The Committee appreciates the opportunity to provide suggestions during the development of this important climate assessment. Attention to the recommendations provided throughout this review report will strengthen the draft NCA4 and improve its ability to reach broad audiences and inform new and continuing adaptation and mitigation responses to climate change.

Chapter 1. Introduction

The Fourth National Climate Assessment is a mandated product developed by the United States Global Change Research Program (USGCRP). The Global Change Research Act of 1990 requires the USGCRP to develop a national climate assessment every four years or less. These assessments are intended to evaluate the state of the science and the broad range of impacts of climate change in the United States. Since the Third National Climate Assessment (NCA3) released in 2014 (Melillo et al., 2014), the USGCRP has moved toward a sustained assessment process. This program evolution provides scientific updates and foundational knowledge that informs the third order draft of the assessment report, "Climate Change Impacts, Risks, and Adaptation in the United States" (hereafter "NCA4"), reviewed in this document. The draft NCA4 report reviewed here serves as Volume II of the assessment. The climate science discussed in the draft NCA4 is based largely on the Climate Science Special Report, or CSSR (USGCRP, 2017), which serves as Volume I of the fourth assessment. The CSSR is summarized in Chapter 2 of the draft NCA4.

The National Academies of Sciences, Engineering, and Medicine convened The Committee to Review the Draft Fourth National Climate Assessment ("The Committee") in November 2017. The Committee is composed of experts able to evaluate the broad range of topics included in the draft NCA4, ranging from climate change science and impacts to responses and risk evaluation. The Committee was specifically charged with addressing the following Statement of Task questions (see also Appendix C for the Statement of Task):

1. Does the report meet the requirements of Section 106 of the Global Change Research Act?
2. Do the key messages reflect current understanding about observed and projected impacts to the United States, the challenges, opportunities and success stories for addressing risk, and identification of emerging issues related to climate change?
3. Does the report accurately reflect the peer-reviewed scientific literature, with a particular focus on literature since the last National Climate Assessment (i.e., since approximately 2013)? Are there any critical content areas missing from the report?
4. Are the findings documented in a consistent, transparent, and credible way?
5. Is the report written at a technical level that is appropriate for the intended audience?
6. Are the report's key messages and graphics clear, internally consistent, and appropriate? Specifically, do they reflect supporting evidence, include an assessment of likelihood, and communicate effectively?
7. Are the data and analyses handled in a consistent, transparent, and credible manner? Are statistical methods applied appropriately?
8. What other significant improvements, if any, might be made in the document?

The Committee held an in-person meeting on November 29-30, 2017, in Washington, DC to discuss the draft NCA4 and begin developing their review report. This in-person meeting also included an open session where the Committee had the opportunity to learn more about the draft NCA4 report and development process from the Director of the National Climate Assessment and Chapter Lead Authors. Additional calls were held to discuss this review report and reach a consensus on the Statement of Task questions. Reviews of each chapter of the draft NCA4 were conducted by small teams of committee members with the relevant expertise who then led discussion of their review with the full committee.

This review report provides a synthesis of the Committee's overarching responses to the Statement of Task for the full draft NCA4 (Chapter 2) and specific comments for the report findings and each individual chapter of the draft report (Chapter 3). General advice and science-based recommendations are also provided for the draft NCA4 Frequently Asked Questions (Appendix A). Detailed line comments on draft NCA4 chapters are provided in Appendix B of this report. The Committee sought to provide constructive criticism that will enhance the draft NCA4, while recognizing that many decisions on report structure and length may constrain considerable expansion of new topics recommended for inclusion.

The National Academies has convened panels to review numerous USGCRP assessments in the past, including the draft Third National Climate Assessment (NRC, 2013; Melillo et al., 2014) and the draft assessment on the impacts of climate change on human health (NRC, 2015; USGCRP, 2016). Most relevant to the NCA4, the National Academies reviewed the draft CSSR (NASEM, 2017b), which was released in 2017 and is referenced throughout the draft NCA4.

Concurrent with the development of this review report, the National Academies directed an independent review of the draft Second State of the Carbon Cycle Report (SOCCR2). The SOCCR2 is a product of the USGCRP sustained assessment process and is intended to provide "a comprehensive assessment of the science and associated human dimensions of the carbon cycle of land, air, and water, with a focus on the United States and North America in a global context." Given the relationship between the draft NCA4 and draft SOCCR2 content, some topics are discussed in both reports. The two committees appointed for these reviews only evaluated the content of the USGCRP draft report that they were appointed to review, addressing the questions in their respective charges. Therefore, some variation in the review of similar topics may be present in the two National Academies' review reports.

Chapter 2. Synthesis of Comments on the Draft Fourth National Climate Assessment

The large body of assessment and primary literature cited in the draft NCA4 provides a strong foundation of science that, when communicated well, can serve as a valuable resource for a wide range of audiences. The Committee found that the most effectively communicated sections of the draft report had three elements: foundational science about the relevant climate change drivers, understandable examples of climate change impacts, and clear examples of adaptation or mitigation actions. The draft NCA4 is generally written at a technical level appropriate for a wide range of stakeholders, but there are many opportunities for the draft report's key messages and supporting information to be conveyed more concisely, with greater cross-referencing between relevant draft NCA4 sections and chapters, and with expanded inclusion of examples of adaptation and mitigation response actions. Additionally, the discussion of uncertainty and risk framing associated with some key messages could be expanded and articulated more clearly. The Committee also recommends identifying the advancements in science and response strategies that have occurred since the publication of the NCA3 to improve the impact of this draft report.

The draft NCA4 builds on the foundational science presented in the CSSR (USGCRP, 2017) to bring the impacts of climate change and response actions into focus for societal decision-making. The CSSR documented the observed 1.0°C (1.8°F) global average temperature increase between 1901 and 2016 that is extremely likely to have been caused by human activities, specifically by the introduction of greenhouse gases into the atmosphere. Beyond increased temperature, the CSSR highlights a wide range of observed changes including the global average rise in sea level of 16-21 cm (7-8 inches) since 1900; the increase in heavy rainfall events, heat waves, and forest fires; the earlier spring snow melt; the decrease in snowpack; the increase in the heat content of the ocean; and associated changes in terrestrial and marine ecosystems, among many other topics. The impacts of these changes in the United States are diverse, complex, and interconnected and are presented in the draft NCA4 around three inter-related themes: national (or sectoral) topics, regional impacts, and responses taken through adaptation and mitigation. Using this structure, the draft NCA4 brings the global and national climate science presented in the CSSR to a scale that is relevant for regional audiences and for groups with a topical focus.

The Committee was impressed by the strength, breadth, and quality of the science presented in the draft NCA4. The draft report provides a thorough and accurate discussion of the predominant aspects of climate change and its impacts, with reasonable reference to the peer-reviewed literature. Specific comments on particular topics and recommended additional citations are provided in the chapter-level comments provided in Chapter 3 and in the line comments included in Appendix B of this review report. A few larger gaps in topical coverage were identified, such as in the draft NCA4 Chapter 7, "Ecosystems, Ecosystem Services, and Biodiversity," and Chapter 29, "Mitigation: Avoiding and Reducing Long-Term Risks," where expanded treatment of some subjects is needed. Minor gaps for draft NCA4 chapters are noted in Chapter 3.

While numerous suggestions on ways to improve the assessment are offered in this review report, these should be viewed as constructive criticism on a generally strong draft report. Furthermore, the Committee believes the draft NCA4 report will be a powerful tool for a variety of stakeholders—including federal agencies, policymakers at all levels (local, state, and national), decision makers, the private sector, community members, interested individuals, educators, and students—to learn about the scope of climate change impacts in the United States as well as possible adaptation and mitigation strategies.

ADDRESSING THE MANDATE IN THE GLOBAL CHANGE RESEARCH ACT OF 1990

The Committee determined that the draft NCA4 meets the intent of Section 106 of the Global Change Research Act (see Box 2.1). The draft report effectively addresses climate change, a critical component of global change, which was found to be an appropriate scope for a national climate assessment. Other aspects of global change, such as land-use change, are introduced with an appropriate level of detail and provide necessary context. Other non-climate stressors that interact with climate (e.g., pollution, aging infrastructure, population growth and associated resource demands) are effectively discussed in some chapters of the draft NCA4 and provide useful context for understanding climate change impacts and societal risks. The Committee encourages greater attention to interacting non-climate stressors when discussing climate change impacts and adaptation strategies where relevant across the draft report. Greater acknowledgement of the limited ability to project future change for some non-climate stressors would also be useful.

In general, the draft NCA4 effectively addresses part 1 of Section 106 in that it integrates, evaluates, and interprets the findings of USGCRP-supported research, including the information synthesized in previously published reports and other data products. It also accurately conveys the scientific understanding of climate change and its impacts to the United States. Evaluation of scientific uncertainties is included in the "Traceable Accounts" section of each draft NCA4 chapter. Suggested improvements related to the uncertainties and traceable accounts content are provided later in this chapter and in Chapter 3 of this review report.

The draft NCA4 also analyzes the effects of global change on the sectors listed in part 2 of Section 106. However, expanded treatment of some of these topics would strengthen the draft NCA4. For instance, the influence of climate change on human social systems should be expanded. Climate change impacts all aspects of society—from food, water, and energy security to transportation and human health—directly and indirectly by amplifying other uncertainties and risks. More explicit discussion of these linkages would help readers to better understand how climate change affects their everyday lives, their region, and the nation.

The treatment of how climate-driven ecosystem change affects ecosystem services and the flow of benefits (or negative impacts) to people needs more discussion. The beneficial role of the natural environment in lessening the impacts of some aspects of climate change should be discussed, such as the value of barrier islands and dunes in reducing effects of sea level rise and storm surges; the role of forests, wetlands, and soils in storing carbon; and the importance of snowpack and natural headwater reservoirs that support downstream water needs. The draft NCA4 could also be improved by highlighting ways in which the natural environment can help to mitigate climate change, such as the replacement of fossil fuel use with renewable energy sources (e.g., solar and wind, hydroelectric, geothermal energy, etc.) and restoration of forests that sequester and store carbon. A comparison of the positive and negative impacts of shifting energy sources could also be discussed (Hill, 2016).

Topics included in the draft NCA4 that are not specifically indicated in Section 106 of the Global Change Research Act provide a valuable addition to the draft report. These include the regional chapters; national topic chapters on air quality, international interests, and interdependences and compounding stressors; and response chapters on adaptation and mitigation. The rich discussions in the regional chapters could strengthen the cohesiveness of the report with increased cross-referencing with the national topic chapters, as appropriate. In the national topic chapters, the Regional Roll-Ups could be better utilized to enhance this added value. The new national-scale chapter on interdependencies could also be used to better connect impacts and responses among sectors and topic areas.

Overall, analysis of current and projected climate trends for the next 25 to 100 years is well integrated throughout the draft NCA4 (Section 106, part 3). Natural climate variability is discussed in the draft NCA4 Chapters 1 and 2, and elsewhere in the report in relation to specific topics, such as El Niño-Southern Oscillation. More discussion of natural climate variability could be added in the regional chapters, particularly for regions where natural climate variability complicates the detection of climate change impacts and adaptation efforts.

LINKING IMPACTS WITH RESPONSE EXAMPLES

The Committee supports the inclusion of response actions and found the examples of these actions in the draft NCA4 to be very impactful. Their inclusion is consistent with literature on science communication, which emphasizes that effective science messaging includes making information relevant to decisions (Moser and Dilling, 2011; Benz et al., 2014; NASEM, 2017a). Discussing climate change impacts alongside examples of current and planned steps to address those impacts, therefore, can leave readers with a sense of how to respond beyond concern or fear of observed and projected impacts.

BOX 2.1
Global Change Research Act, Section 106. Scientific Assessment

On a periodic basis (not less frequently than every 4 years), the Council, through the Committee, shall prepare and submit to the President and the Congress an assessment which:

1. Integrates, evaluates, and interprets the findings of the Program and discusses the scientific uncertainties associated with such findings.
2. Analyzes the effects of global change on the natural environment, agriculture, energy production and use, land and water resources, transportation, human health and welfare, human social systems, and biological diversity.
3. Analyzes current trends in global change, both human-induced and natural, and projects major trends for the subsequent 25 to 100 years.

Figure 1.5 (also included as Figure 29.1) was identified as a particularly effective graphic because it shows actions at both local and state levels in a straightforward manner and represents most of the regions included in the draft NCA4.

Because of the effectiveness of response examples, greater reference to these activities is recommended where such examples exist. Adaptation is discussed in five of the ten regional chapters and mitigation in only three. Since the NCA3 was published, many new response actions have been implemented and planned. Highlighting these actions provides an opportunity to demonstrate advancement in recent years. In particular, recognition of actions that have occurred in the private sector or through public-private partnerships should be acknowledged in addition to government-led efforts.

Finally, given the importance of adaptation and mitigation and the recommendation that these topics be better emphasized throughout the draft NCA4, the Committee suggests making the adaptation (Chapters 28 in the draft NCA4) and mitigation (Chapter 29) chapters more prominent in the report by placing them earlier in the chapter list. In such a long draft document, having the response chapters at the very end may give the misperception that they are less important or not discussed across the draft report.

Recommendation: Incorporate more examples in the draft NCA4 that highlight new and ongoing adaptation and mitigation activities. These should include actions in the private sector, public-private partnerships, and government at multiple scales.

COMMUNICATING REPORT FINDINGS

The Overview Chapter (Chapter 1) of the draft NCA4 was reviewed with particular attention to whether the chapter provides an appropriate and balanced overview of the draft report content and is written at a technical level appropriate for the intended audience. As an overview chapter, Chapter 1 should serve as a go-to for readers interested in gaining a quick understanding of the NCA4 and complement the "Report in Brief" that will be developed by the NCA4 authors. While the chapter is well written and scientifically accurate, the strong emphasis on climate science is out of balance with the impacts focus of the draft NCA4 as a whole. Revising this chapter to focus around the twelve report findings and national topic and regional impacts would strengthen the chapter.

Recommendation: Reframe the Overview Chapter of the draft NCA4 to center around the twelve report findings that reflect the impacts and responses that are discussed throughout the draft report.

COMMUNICATING KEY MESSAGES

Key messages in the draft NCA4 convey relevant information and are well supported by the peer-reviewed literature, but most are long and contain multiple unique points. The Committee recognizes that the NCA4 authors were given instructions to limit the number of key messages and this objective may have led to the complexity of some individual key messages. However, using more concise language and shortening the key messages would make them more impactful. Messages could be prioritized by focusing more directly on the take-home point and listing this first. Key messages could also be shortened to provide only the portions of the messages where confidence and likelihood statements are provided in the traceable accounts. Making the length of key messages more consistent across the report is also recommended and where appropriate, key messages in national topic chapters should parallel related key messages in regional chapters in order to improve consistency and linkages across the draft NCA4.

Many of the key messages place stronger focus on climate change itself rather than the subsequent impacts of that change or associated response actions. A more balanced treatment of these key messages is suggested in order to place greater emphasis on impacts and responses that are the primary focus of the draft NCA4. The NCA4 authors should also consider strengthening the emphasis on adaptation and mitigation activities associated with key messages as appropriate. This modification will facilitate more solution-oriented messaging, which can be lost when key messages are framed largely around the climate drivers and impacts.

Key messages and supporting text are most compelling when concrete examples are given, including stories where response actions have been shown to reduce climate change impacts. The Committee found the draft NCA4 Chapter 18, "Northeast," to be a good example of effective key message use and thinks it could serve as a model for other chapters. In Chapter 24, "Northwest," the framing of vulnerable communities as being on the front lines of climate change was viewed as excellent. It is a way to draw in diverse communities that will be impacted by climate change, with consideration for socioeconomic factors as well as location, livelihoods, cultural practices, distance to services, etc. The Committee recommends that this language be used throughout the report, where appropriate, as it greatly increases the relevance of key messages for diverse audiences.

Recommendation: Key messages should be presented using more explicit and concise language. Examples that align with key messages should be included wherever possible in the supporting text and figures. More of the key messages should be supported by examples of response actions to facilitate solution-oriented communication and information sharing.

Use of Graphics to Support Key Messages

Graphics in the draft NCA4 were found to be a mix of highly informative material that is well aligned with the key messages and figures that are difficult to understand and not well tied to chapter content. It is recommended that the NCA4 authors include figures that align closely with chapter key messages. The Committee encourages identifying opportunities to use similar graphics in multiple chapters (i.e., linking concepts in topic chapters to examples in regional chapters) in order to improve message consistency. In general, graphics that convey enough information on the figure and require only limited caption text are the most effective and allow readers to understand take-away messages easily.

Traceable Accounts to Support Key Messages

The Traceable Accounts section included in each chapter is a critical component of the draft NCA4 and a challenging section to review. This section is where the NCA4 authors use their expert judgement to assign confidence and likelihood based on the available evidence contained in the literature. To rigorously review the confidence and likelihood determinations made by the NCA4 authors would require an expansive panel of experts with expertise in all of the detailed topics included in the draft NCA4, as well as a thorough understanding of how author judgements were developed. The Committee's review of the traceable accounts focused largely on whether the NCA4 authors presented an adequate level of detail about the information they used in identifying the content of key messages and determining the appropriate confidence and likelihood designations that they assigned.

Most key messages contain an assessment of confidence and not likelihood (see definitions on page 8 of the draft NCA4), which the Committee deemed appropriate given the generally qualitative nature of the key message content. Minor chapter-specific concerns about confidence and likelihood are provided in Chapter 3 of this review report. Many chapters effectively utilize traceable accounts and provide robust support for key messages presented in a transparent manner, but some inconsistency in the utilization of this section was identified. For instance, some chapters use the traceable accounts largely to detail *how* the evidence was obtained while the traceable accounts in other chapters focus on an explanation of *what* evidence supports the key messages. According to the Front Matter "Guide to the Report" section of the draft NCA4, both of these types of information should be provided. In other cases, the traceable accounts introduce new information not previously discussed in the chapter. One of the chapters identified as being particularly effective in developing traceable accounts was the draft NCA4 Chapters 21, "Midwest."

The traceable accounts section is also an appropriate place to provide sufficient evaluation of topics that are introduced in the main text, but where the scientific evidence is contrary or mixed and where different conclusions could be reached (e.g., changes in likelihood or confidence since the NCA3 or key messages with low or medium confidence). This section could also be used to expand on topics that NCA4 authors deem necessary to include in the report but where confidence is currently low based on available evidence, such as impacts of high consequence where there is currently low confidence.

Other Comments About Key Messages

Broader discussion of equity and environmental justice is needed in the key messages and supporting text in the draft NCA4. Currently, this topic receives very little treatment except in the draft NCA4 Chapter 8, "Coastal Effects," (most notably Figure 8.4). There is literature that could be drawn on to support wider inclusion of this topic in the draft report (e.g., Brulle and Pellow, 2006; Balazs et al., 2012; Bautista et al., 2015; Bravo et al., 2016).

The Statement of Task charges the Committee with evaluating statistical methods applied to the key messages and supporting text in the draft NCA4. The majority of the information in the draft report is an evaluation of previously published literature and therefore does not contain new analyses or statistical tests. Comments related to analyses for a few specific draft NCA4 chapters are provided in Chapter 3 of this review report.

Note that any revision of key messages would require revision of the main text, traceable accounts, and report findings in order to assure consistency with the revised messages across the draft NCA4 report.

COMMUNICATING UNCERTAINTY AND RISK

The draft NCA4 report deals with a broad range of climate change impacts that can influence societal risks. These impacts carry uncertainties that arise from predictive capabilities, environmental stochasticity, feedbacks, interactions among climate change impacts and non-climate stressors, and many other factors. By evaluating both the magnitude of potential impacts and the probability of occurrence, while considering known characterized uncertainties, the draft NCA4 can support risk-informed decision-making.

The NCA4 demonstrates an improvement compared to the NCA3 in its treatment and communication of uncertainty and the use of a "risk framing" structure. Steps were made early in the NCA4 development process to inform this advancement, including the convening of a National Academies workshop, "Characterizing Risk in Climate Change Assessments" (NASEM, 2016). This workshop was intended to advance the conversation on "characterizing the risks and clearly framing them in terms of their implications for people and systems," "conveying clear and accurate information about those risks in ways that are useful and accessible," and "identifying the connections across sectors and regions that are critical for understanding risks."

The Committee found the Call Out Box "Why is Risk Framing a Useful Tool for Decision-Makers" in Chapter 1 (pages 48-49 of the draft NCA4) to be useful in explaining the complexity of evaluating uncertainties and risks associated with interconnected systems. Importantly, it also notes that decisions can be (and are) made in the presence of uncertainty. However, an explicit distinction between "risk" and "risk framing" is needed. The term "risk" is not defined in the draft NCA4, such as risk due to potential loss from adverse events as used in engineering or the impacts of uncertainties typically used in other sectors including financial markets. Risk seems to be used more colloquially rather than to provide a formal quantification of risk (i.e., the cross product of the magnitude of an impact/consequence and the probability of occurrence), yet the quantification of risk is listed as an element of risk framing (see "Treatment of Uncertainties: Risk Framing, Confidence, and Likelihood" on page 7 of the draft NCA4). Perhaps as a result of this ambiguity, the individual chapters in the report discuss risk in a somewhat inconsistent fashion. The many climate change impacts, sources of uncertainty, and risks discussed in the draft NCA4 need to be differentiated more clearly. Different types of risk call for different types of actions and risk management solutions. Greater standardization in the treatment of this issue across the draft report is recommended. Similar types of uncertainties and risks should be discussed across chapters and each type should be clearly articulated. Uncertainties and risks that may be unique to an individual sector(s) could also be defined in clear terms and discussed.

Although uncertainty and risk concerns are noted by the Committee, some chapters and topics were handled well in the draft NCA4. Chapter 28, "Near-Term Adaptation Needs and Increased Resiliency," provides an appropriate template for risk framing through robust reporting on major uncertainties, confidence, and likelihood and could provide an example for other chapters. Additionally, a well-articulated acknowledgement of sectoral interdependencies and their effects on uncertainty and risk, multiple stressors, and complex systems is a primary component of the draft report—although this topic is largely relegated to Chapter 17 rather than embedded more broadly throughout the draft report. Interconnectedness and interdependencies spatially and across sectors increase uncertainty that could lead to negative, cascading impacts. In contrast, spatial interactions among sector activities can also reduce impacts in cases where synergies are positive or countervailing. The NCA4 authors should consider acknowledging these nuances more extensively in the report.

Additional comments related to the treatment of uncertainty and risk in the draft NCA4 are given in the review of the draft NCA4 Chapter 1 (see Chapter 3 of this review report).

Recommendation: The various types of risk included in the draft NCA4 should be defined more explicitly. Improved consistency in the types of risk discussed and inclusion across chapters is also recommended.

BRIDGING TOPICS AND SCALES

The Committee found the linkages and cross-referencing to relevant information among chapters to be insufficient. In such a large report, it is essential to guide readers through the information in a manner that will allow them to explore their topics of interest, making the report more broadly useful. Many of the national topic chapters are siloed such that it is challenging to find the relevant connections to related issues in other chapters. As understanding of the interconnected nature of climate change impacts and responses improves, the need to make robust connections among topics becomes more important. The traditional structure of and process for developing the NCA reports—particularly the topic/sectoral chapters—may hamper the ability to explore interconnections and interdependencies in a way that captures the complexity of the issues. The national topic chapters have a relatively narrow scope and are authored by relevant experts on the subject matter. Thus, finding intersections is not their primary focus, but rather a secondary consideration. The inclusion of the new chapter on "Interdependencies, Multiple Stressors and Complex Systems" (Chapter 17 in the draft NCA4) begins to address this shift; however, as a stand-alone chapter, the concepts are not well integrated across the draft report.

The "Regional Roll-Up" sections in the national topic chapters demonstrate an effort by the NCA4 authors to bridge chapters. However, missing linkages and inconsistencies in the treatment of topics are evident, which could send mixed messages to readers. For example, the topic of ecosystems shows up in the key messages for each region with a wide range of highlighted impacts. The focus of Chapter 7, "Ecosystems, Ecosystem Services, and Biodiversity," is narrow, with strong emphasis on invasive species, and does not capture the range of impacts on ecosystems or ecosystem services spanning from the U.S. Caribbean to Alaska. Similarly, the issue of water (Chapter 3 in the draft NCA4) shows up in the key messages in many regions, but the water chapter focuses largely on the impact of changing temperatures on snowpack and runoff while the regional chapters emphasize the timing of precipitation and increasing extreme events as important impacts.

Generally, the topics are connected well within the individual regional chapters, in large part because they provide a relatively holistic treatment of relevant climate change impacts and they discuss interrelated issues together. This approach logically conveys the complexity of impacts that extend across sectors/topic areas.

As a whole, the draft NCA4 lacks a unifying figure that ties together the regional information and highlights the largest impacts, risks, and responses. While it is impractical to think that the entire contents of the draft NCA4 can be condensed into a single figure, it is suggested that the NCA4 authors consider dominant impacts and response activities unique to different regions and explore how these could be illustrated in a single figure or set of figures. A graphic of this nature presented in the Overview Chapter could be highly effective in communicating the high-level findings of the NCA4. The Committee strongly recommends working with a graphic artist to develop such a product.

Recommendation: Linkages to interrelated topics among chapters should be increased throughout the draft NCA4 to ensure consistent treatment of similar topics and to provide readers with a clearer understanding of how impacts and responses at national to regional scales are connected.

HIGHLIGHTING NEW DEVELOPMENTS IN CLIMATE SCIENCE, IMPACTS, AND RESPONSES

The Committee recommends more clearly distinguishing between what information is new since the publication of the NCA3. Currently in the draft NCA4, citation of new material is often embedded with older references in the text. In situations where new literature provides further support for a finding discussed in the NCA3, this may be appropriate. However, in instances where significant advancements in the science have been made or where new actions or activities have been recently implemented or planned, new information should be identified more explicitly. An example of this is for the Northeast region, where the NCA3 identified adaptation plans but little action could be reported. Now, there is action in most states that is discussed in the draft report. Providing greater emphasis on new knowledge and actions may also facilitate more solution-oriented messaging across the report, as recommended in the "Communicating Key Messages" section found earlier in this chapter of the review report.

A high-level overview of new developments could be included in Chapter 1, "Overview," of the NCA4. This approach was effective in the CSSR (see Box 1.2, "A Summary of Advances Since NCA3," USGCRP, 2017). For individual chapters, some possible options for distinguishing this new material would be to color code the key messages or provide notation or language that explains whether a key message has changed since the NCA3 based on new scientific understanding or may be the same or similar to the NCA3 but has new evidence that provides continued support. New key messages could begin with language such as "new evidence [either] suggests a new conclusion be discussed or confirms prior understanding." If the confidence, likelihood, or conclusions about any key messages have changed since the NCA3, that too should be explained. Alternatively, each chapter could select a few areas where significant progress has been made or where a change in confidence or likelihood has occurred between the third and fourth assessments and focus a portion of the chapter discussion more prominently on those areas, with appropriate context.

A brief description of the relationship between materials included in the NCA3 and the NCA4 would be a beneficial addition to the Front Matter of the draft NCA4. In the spirit of the iterative and sustained assessment process, it is assumed that the NCA4 builds on the content in NCA3, but how this transfer and updating of information was done could be made more transparent. This would provide clarification about the selection of materials that are included in the draft NCA4 that may (or may not) have been in the NCA3. For instance, are there key messages in the draft NCA4 that did not appear in the NCA3 due to a change in the science or a shift in the prioritization of topics?

The draft NCA4 would also benefit by more thoroughly identifying recently-developed state and local climate assessments and adaptation planning efforts, as well as other available climate tools. If appropriate, relevant results from these sources could also be highlighted. An indication of which efforts provide actionable information could also be stated. In general, it seems that the number, depth, and breadth of such assessments has increased substantially since the publication of the NCA3; if so, that aspect should be clearly stated. Additionally, it could be noted that more climate adaptation planning has begun to occur at the community level—in some cases, it was ordained by state legislation, and in other cases, it was by local policy or interest. If possible, linking these resources and planning examples to the draft NCA4 would allow the report to serve as a broader resource for the NCA4 audience. This connection would be particularly effective in the regional chapters. Reference to an expanded list, where possible, of available regional-scale climate projections and assessments would improve comprehensiveness. The Committee also suggests drawing more on relevant gray literature (when it is found to meet the NCA4 information quality guidelines) to provide a more comprehensive evaluation of the state of knowledge and to draw on examples of local and regional adaptation and mitigation strategies. Citations taken from media sources should be avoided; instead, reference should be made to the primary sources.

Recommendation: Authors of the draft NCA4 should explicitly identify significant advancements made since the Third National Climate Assessment, with emphasis on emerging science, impacts, and examples of new response actions.

Chapter 3. Comments on Each Chapter of the Draft Fourth National Climate Assessment

FRONT MATTER: REPORT FINDINGS

The Committee reviewed the twelve report findings with particular attention to whether the findings effectively synthesize the key messages and link information provided throughout the draft NCA4 report in a logical and consistent manner. Additionally, the Committee evaluated whether the report section is communicated effectively and the level of technical detail is appropriate for the intended audience.

The report findings are well written and for the most part, they reflect the topics emphasized in the draft NCA4. However, the linkages made in the findings are not always present across individual chapters; some important topics that receive emphasis in chapters are not adequately reflected in the report findings, and in some instances, information provided in the report findings is not well supported in the draft NCA4 chapters. The effectiveness of communicating the report findings may also be improved if some findings are discussed earlier in the findings list, to draw more fully upon the strengths of the draft NCA4. Specific recommendations for addressing these concerns and strengthening the impact of the report findings are provided in this section.

Comments on Select Report Findings

> **1. Communities.** Climate change exacerbates existing vulnerabilities in communities across the United States, jeopardizing people, economic growth, and quality of life and creates new risks that are projected to intensify without adaptation and mitigation.
>
> **10. Indigenous peoples.** Climate change increasingly threatens tribal and Indigenous communities' livelihoods, economies, health, and cultural identities through disruption of interconnected social, physical, and ecological systems.

The Communities and Indigenous Peoples Report Findings (and supporting text) effectively link society to a range of climate change impacts discussed in the draft NCA4, which should help messages resonate with readers. The Committee suggests that the NCA4 authors consider broadening the Indigenous Peoples Finding to also include vulnerable populations, since vulnerable populations (i.e., Indigenous Peoples and other vulnerable populations) are discussed in many chapters and may experience many similar impacts and challenges associated with climate change. The discussion of vulnerable populations is especially effective in draft NCA4 Chapter 24, "Northwest," which could be used as a model to revise this text. Presenting the Communities and Indigenous Peoples Report Findings side-by-side early in the list of findings should also be considered given their similarity and many overlapping impacts.

> **12. Adaptation and Mitigation.** Communities and businesses are working to reduce climate change-related risks and their associated costs through the adoption of robust and proactive management and adaptation strategies that are viable for a wide range of climate futures.

The Adaptation and Mitigation Report Finding would benefit from some mention of how much adaptation is being implemented and whether there are examples that demonstrate a measurable reduction in climate change impacts as a result of implemented actions (i.e., successes). As detailed in Chapter 2 of this review report, the Committee found examples of response actions to be highly effective for communicating impacts, and discussing this topic more prominently would benefit the report. It would be appropriate to succinctly convey a similar message for this report finding. Listing this finding earlier in the list of findings could also be considered, to give more prominence to response actions. Finally, the Adaptation and Mitigation Report Finding focuses largely on adaptation with little mention of mitigation. It is suggested that better balance be provided.

> **11. Interconnected Impacts.** As climate risks intensify, the interdependent systems on which we rely are vulnerable to cascading impacts across sectors, threatening essential services and sectors within and beyond the nation's borders.

The Interconnected Impacts Report Finding generally conveys a message applicable to other findings: impacts are linked to one another in complex ways that cannot be sufficiently understood in isolation of one another. The draft NCA4 Chapter 17, which focused on interdependencies, is a strong chapter and its messages should be more broadly referenced across the draft NCA4. That strength (and broader representation across the report) should be captured in the report findings. Placing this synthetic finding earlier in the report findings list could facilitate this and may provide additional context to readers as they read through the full report findings list. Additionally, the Interconnected Impacts Report Finding should be modified to include social systems, including communities. The interconnected nature of climate risk is an issue across almost every topic identified in this report and the societal component is an essential piece of that linkage.

> **2. Economy.** Losses to infrastructure, property, and productivity driven by the impacts of climate change are expected to increasingly disrupt the U.S. economy, even with mitigation and adaptation efforts.

The Committee found the Economy Report Finding to be appropriate and important to include in the draft NCA4. However, the draft report chapters provide little support for inclusion of this finding. Treatment of economic impacts of climate change is very uneven; this topic is constrained largely to Chapter 29, "Mitigation: Avoiding and Reducing Long-Term Risks," and is only briefly discussed in other places. This may be understandable because defensible economic impact studies are limited and not available for all topics. Generally, a more coherent strategy for handling this topic is needed. It is recommended that discussion of economic impacts of climate change be better woven throughout the draft report where there is available literature to support it and where information is not available, to explicitly say so. The Economy Report Finding should also include language that constrains the framing to areas where robust research exists so that the finding is not misinterpreted as encompassing the entire U.S. economy.

Recommended Topic Additions and Linkages

- **Energy:** Expanded discussion of energy in the Economy and Interconnected Impacts Findings would be beneficial, in addition to its current inclusion in the Infrastructure Finding. Energy is a critical system that is discussed in many draft NCA4 chapters. It supports economic activity and serves as an integral component of interconnected infrastructure systems that are vulnerable to climate change impacts.

- **Wildfire:** Wildfire poses large risks relevant to the Ecosystems and Ecosystem Services, Agriculture, and Communities Report Findings and is discussed in many national topic and regional chapters. Inclusion of wildfire in the findings would also likely resonate with NCA4 readers given recent trends in wildfire activity in the United States.
- **Biodiversity and Ecosystem Services:** The interaction and fundamental underpinning of biodiversity and ecosystem services (draft NCA4 Chapter 7) is covered in many other draft NCA4 topic chapters (e.g., Water, Energy, Land Cover and Land Use Change, Forests, Coastal Effects, Oceans and Marine Resources, etc.) and should be highlighted more fully in the Ecosystems and Ecosystem Services Report Finding.
- **Increased Topic Linkages:** Other linkages that would strengthen the report findings include mentioning forestry in the Agriculture Finding and including ocean ecosystem adaptation and mitigation actions in the Tourism and Recreation Finding.
- **Increased Regional Linkages:** The Committee suggests considering the value of incorporating regional context into the report findings. While the report findings cover broad topics, the regional chapters demonstrate that impacts are variable across the nation and that some of the report findings will resonate more strongly in some regions than in others.

Other Report Finding Comments

In a few cases, the statements in report findings are either inconsistent or not well supported by chapter text. For instance, the report findings emphasize Indigenous vulnerability to a greater extent than does the chapter on that topic. For the Agriculture Finding, the statement "economies of agricultural regions at risk" is not well-developed in Chapter 10, "Agriculture and Rural Communities." Careful review to ensure that report finding content is supported by chapter text and that similar emphasis is placed on the topics covered is needed.

A few national topic chapters of the draft NCA4 were not strongly captured in the report findings and the Committee leaves it to the NCA4 authors to determine whether this is an appropriate omission. These chapters include Chapter 5, "Land Cover and Land Use Change," and Chapter 16, "Climate Effects on U.S. International Interests."

Additional revisions to the report findings should also be made as appropriate, if changes to associated key messages occur as a result of other recommendations provided in this review report.

CHAPTER 1: OVERVIEW

Summary

The Committee found the material contained in Chapter 1 to be accurate and well written, with attention given to many of the aspects of climate change that are dominant in the discussion of impacts including sea level rise, temperature, and precipitation. However, the strong emphasis on climate science in this chapter is out of balance with the impacts focus of the draft NCA4 as a whole. Broadly, the Committee recommends placing greater emphasis on the impacts and response efforts in Chapter 1, including in the selection of figures. Expanded synthesis of the national topic and regional chapter contents, as well as some attention to new science published since the NCA3, is also needed.

Chapter 1 should be the synthesis or bridge from the nationwide topic issues and the regional analysis to the major report findings. While the Committee supports the draft NCA4 report findings, this synthesis across topics and scales is lacking. The Committee recommends revising Chapter 1 using the framework of the twelve report findings, with impacts from the national topic chapters and regional analysis being the focus.

Treatment of Uncertainty and Risk

As noted in Chapter 2 of this review report, the content of the Call Out Box "Why is Risk Framing a Useful Tool for Decision-Makers" is helpful to readers. Importantly, it discusses the use of risk framing as a means to convey information in a way that may inform decisions about responding to impacts in the presence of uncertainty. Using complex systems and interdependencies to explain these concepts is effective. The Call Out Box could also be a good place to point out that risk framing is an appropriate framework to discuss both adaptation and mitigation and to talk more about better (adaptive) risk management and point to those alternative approaches. The draft report generally does well in citing and communicating the recent research on the treatment of risk. However, some recent papers could be added, including Bakker et al. (2017), Oppenheimer et al. (2016), and Wong and Keller (2017). Information provided by the International Organization of Standardizations could also serve as a useful resource for further exploring risk (ISO, 2009).

Chapter 1 would be a good place to elaborate on the different types and drivers of uncertainty and risk and how to manage them. Alternatively, some of the material on how to manage risks could be mentioned in the Call Out Box "Confidence and Uncertainty in Climate Science" or in Section 1.4 or Section 1.6. For example, in the discussion of shortcomings of cost-benefit analysis, better alternatives could be introduced, including robust decision-making (citing, for example, Hall et al., 2012; Herman et al., 2015; Ranger et al., 2013; Lawrence et al., 2018) and risk-risk analysis (Viscusi, 1994). Expanded discussion of economic impacts of climate change in the context of risk for more sectors/chapters would also be welcome, as discussed in the context of the Economy Report Finding in the "Front Matter: Report Findings" section earlier in this review chapter.

Chapter 1 would also be the appropriate place to clearly define the scope of uncertainty and risks evaluated in the draft NCA4, preview different types and sources of uncertainty and risk in climate assessment more generally, and discuss how to manage them. As described in Chapter 2 of this review report, it is important to differentiate between sources of uncertainty and risk because different types of risk call for different types of risk management solutions. While the draft NCA4 report cannot be expected to go into detail on all these aspects of uncertainty and risk, it would be helpful if Chapter 1 could provide some conceptual guidance or advice on performing risk analysis and management in a way that accounts for how impacts interact across sectors and scales (e.g., Haimes, 2009; Ayyub, 2014) as well as a generalized treatment of uncertainty (e.g., Klir, 2005; Ayyub and Klir, 2006).

More specifically, discussion of uncertainties and their effects on evaluating risk should differentiate between the following:

- Negative consequences/impacts imposed by climate change
- Uncertainty introduced in predicting events and outcomes due to climate change (as the future is no longer like the past) and increased climate variability, and sources of uncertainty in these predictions
- Effects of the stochastic nature of the environment
- Compounded events
- Model-based uncertainties
- Uncertainties due to divergent expert opinion
- Effects of knowledge gaps due to non-existing research
- Feedbacks and interaction between different sectoral risks
- Feedbacks and interactions between infrastructure (built systems), ecosystems (natural systems), and social systems (interconnected impacts)
- Feedbacks and interactions between climate risks and non-climate stressors (e.g., aging infrastructure, stressed ecosystems, social inequality), which are mentioned on page 19 of the draft NCA4
- Risk of underutilizing existing climate and climate change forecasts, willfully or due to individual or institutional myopia and status-quo bias
- Risk of not having adequate theory and tools to model and predict complex emergent risks

Specific Comments on Chapter Content

In order to help bolster the impacts discussion in Chapter 1, it is suggested that the treatment of adaptation be expanded in Section 1.6. Considerable development of new adaptation activities since NCA3 has occurred and these actions should generally be discussed more thoroughly in Chapter 1 of the draft report. The Committee recognizes that adaptation often occurs at local to regional scales, so it would be difficult to provide a comprehensive national-scale summary in Chapter 1, but a table or figure that provides some context for recent growth in this area is warranted. This would also contribute to improved distinction between the NCA3 and NCA4.

The Call Out Box "What's New in NCA4?" lists new data products, report chapters, and framing of risk and economic impacts. While this nicely summarizes the advancements in the processes and information available for NCA4, the title could be interpreted to mean that the box discusses new advancements in scientific understanding or what topics may be new in NCA4 relative to NCA3. The Committee recommends providing a clearer distinction about what is new in the science and response action in the draft NCA4, possibly in this Call Out Box. See also "Highlighting New Developments in Climate Science, Impacts, and Responses" in Chapter 2 of this review report for recommendations on approaches to making this distinction.

For the Call Out Box "Confidence and Uncertainty in Climate Science," it is suggested that "confidence" be defined so that it is not misinterpreted in this context. Additionally, the meaning of "increasing confidence in climate science" may not be clear and could be misinterpreted—is this asserting that the public has increasing confidence in the climate science community? The Committee thinks this probably means to say that because observed signals—and, to some extent, projections—have become stronger, there is increasing confidence by scientists in the results that have accumulated since NCA3, which reinforces conclusions and climate change trends that were described in NCA3. The language should be updated to better convey this message, or the message intended by the Chapter 1 authors.

The draft NCA4 would benefit from a unifying graphic (or a few graphics) that distills key messages from the national topic and regional chapters. Chapter 1 is an appropriate location for such a graphic. The Committee suggests identifying a dominant climate change impact in each region and an example of adaptation or mitigation that addresses that impact (ideally something that can be summarized well visually). Overlying this information on a national-scale map with the NCA4 regions identified could serve to focus the draft report material in a new way that also highlights response efforts. An approach to thinking about this unifying graphic would be to ask: "What are the fundamental figures that readers should remember from the NCA4?" The Committee recommends that the unifying graphic be accompanied by a short paragraph in Chapter 1 emphasizing the inter-connections among regions, sectors, and biophysical and social components of ecosystems. The structure of the NCA4 report requires separate treatment of such topics in order to provide necessary detail, but the integrated nature of climate change impacts and strategies to address them should not be understated.

The Committee would like to see the draft NCA4 expand discussion of impacts and risks imposed by climate change on National Parks (see Monahan and Fisichelli, 2014). It is appreciated that the draft NCA4 Chapter 1 mentions the impact of particulates from wildfires on the scenic vistas in National Parks (page 48) and one of the Frequently Asked Questions addresses melting glaciers in Glacier National Park. However, there is relatively little mention in the national topic or regional chapters of the much broader suite of likely climate change impacts. This might be expected to be included in the draft NCA4 Chapter 7, "Ecosystems, Ecosystem Services, and Biodiversity." Placing greater emphasis on the impacts and risks to national parks, which are popularly described as "America's Best Idea," could be an effective outreach to the NCA4 readers, since national parks are iconic and representative of the impacts that will be experienced in state parks and other public recreational areas. As noted in their mission statement, "The National Park Service preserves unimpaired the natural and cultural resources and values of the National Park System for the enjoyment, education, and inspiration of this and future generations" (although recognizing that "preserving unimpaired" must be tempered by the fact that environmental and social change may prohibit preservation or restoration to conditions during pre-European times [Revisiting Leopold: Resource Stewardship in the National Parks, 2012]). This mission statement strongly espouses conservation and education over a sustained period.

It is recommended that the full NCA4 author team carefully review the contents of the draft NCA4 Chapter 1 to ensure that the information included is well justified by the associated national topic and regional chapters. In a few cases, the Committee identified instances where the underlying chapter did not adequately support the information included in Chapter 1. For example, the negative impacts on timber prices and economic well-being of forest landowners (page 36, lines 16-20) are not discussed specifically in the forests chapter (Chapter 6 in the draft report), except by reference to a single paper. Similarly, the effects of ground-level air pollutants (page 44, line 7) is discussed in the context of forests, but this topic is not discussed in Chapter 6 except by a single reference; where mentioned, the influence of ground-level ozone is not well explained (see the review for Chapter 6 for more detail).

In general, the regional chapter impact messages are not described clearly in the draft Chapter 1. Inclusion of a unifying graphic, as mentioned earlier in this section, could be one approach to addressing this shortcoming.

Comments on Graphics

The Committee determined that the figures included in the draft NCA4 Chapter 1 are accurate and well selected for demonstrating the state of climate science. However, as noted previously, it is recommended that the authors add a unifying graphic and place more focus on responses and impacts in order to better balance the chapter content with the draft NCA4 as a whole. Comments here are recommendations for the figures contained in the draft NCA4, which the Committee recommends editing if retained.

Figure 1.1

The quantity of information in Figure 1.1 is overwhelming, often overly complicated for the intended audience, and lacking in explanation of some important details. With so much data contained in a single figure, it is challenging to identify what key message(s) the graphic as a whole is trying to convey, and what take-home messages are intended. The text referencing the figure is spread throughout the draft chapter, making it difficult for the figure to stand alone. However, the content was determined to be generally important, and this section outlines a number of suggestions to simplify the graphic and/or draw out the content more directly.

The content of Figure 1.1 is closely aligned with the physical climate science detailed in the CSSR and summarized in Chapter 2 of the draft NCA4. The information is accurate, but careful attention should be given to which portions of the figure are necessary and effective in conveying the climate change impacts that are the central focus of the draft NCA4. Some impacts may not resonate or be readily understandable to a general audience without more detail (e.g., snowpack).

Nearly every time series presented is for a different set of years. The inconsistency in time periods shown is confusing and makes it difficult to readily compare changes across panels (e.g., panel (h) shows area burned by wildfire from 1983 to 2016, while panel (i), located just below (h), shows the percent of U.S. area experiencing drought from 2000 to 2016). Time periods should be standardized wherever possible to make the figure panels more consistent. Additionally, the same colors are used to mean different things in different panels, which may confuse readers (e.g., in panels (a) and (b) red depicts an increase, and in (d) it represents a decrease).

The caption could be better utilized to explain Figure 1.1 and orient the reader. It is unclear what the arrows located on the left side of individual panels are intended to show. Terms like snowpack should also be defined.

An infographic that complements Figure 1.1 could be an approach to simplify the main messages and make the material more accessible to the intended audience. This infographic could include just the arrows provided on the panels or some other tangible element.

It is unclear whether the Figure 1.1 panels are shown elsewhere in the report. Since this figure is presented in the draft NCA4 Overview Chapter, the Committee expected the data to be more fully explained elsewhere in the report; if this is not the case, it may be misleading to readers.

Given the small size of individual panels in Figure 1.1, they are difficult to read. To address this, the figure could be separated into four figures that individually explain the four primary topic areas: Weather and Climate, Snow and Ice, Land and Water, and Oceans and Coasts.

Figure 1.2

Figure 1.2 accurately summarizes information provided in Chapter 2 of the draft NCA4, "Our Changing Planet." The figure also nicely illustrates the evidence that supports the conclusion that human factors are the dominant influence on recent climate change. However, the figure is relatively complicated for a general audience. To address this, the Committee suggests expanding the caption to more fully summarize the contents in language appropriate for a broad audience. Cross-reference to Figure 2.1, which shows the separate effect of each natural forcing (solar and volcano), combined human-caused forcing, and natural variability on global mean temperature, could be useful for readers interested in individual causes of temperature change. The Chapter 1 authors could also consider developing a graphic more similar to Figure 2.1 to make the messaging of the report and the graphics consistent across chapters.

Figure 1.3

It is recommended that the legend showing the Representative Concentration Pathways (RCPs) be expanded to also include language indicating relative emissions (e.g., lower scenario, higher scenario, etc.) to be easily understood by the intended audience and consistent with the caption.

The use of different baselines for the projected temperature change and the historical observed temperature change may be confusing for some readers. If a common frame of reference could be applied, it is recommended; otherwise, additional explanation of the difference should be provided. Also, the y-axis title indicates the graphic is showing temperature change relative to a 1986-2015 baseline while the caption lists the baseline as 1986-2005, suggesting that one of these years is a typo and should be corrected. This also applies to where this graphic is included as Figure 2.2 in the draft NCA4. It is also suggested that the graphic be made larger (similar width to Figure 1.4) and that the authors consider listing the legend information to the right of the figure so that the RCP/scenarios information is listed to the right of the y-axis showing temperature change projected for the end of the century.

Figure 1.4

Figure 1.4 conveys too many messages for the intended audience to easily follow. In particular, the relationship between the projection scenarios on the graph and the RCPs shown to the right of the figure may not be clear to a non-technical audience. A figure with this level of detail is appropriate for Chapter 2 of the draft NCA4, where it is also provided as Figure 2.3, but should be simplified if retained in Chapter 1. This could be achieved by showing fewer scenarios and/or making the graphic more visually consistent with Figure 1.3. It is also suggested that the height of the figure be increased so it is more legible and that the y-axis showing feet be converted to inches so that any differences in sea level rise projected in the near term may be visible. The use of both feet and meters on the y-axis but description in text of only feet may be confusing to some readers. It is also recommended that the time period shown on the x-axis be shortened to focus more on the projections rather than the historical rate of rise, or that more justification for the inclusion of the long time period be provided. The Chapter 1 NCA4 authors could consider adding a graphic of a person of average height next to the right side of the y-axis as a way to make the impact of the amount of projected sea level rise more tangible. Finally, it is suggested that the sea level rise graphic used in this chapter show national estimates of sea level rise instead of global estimates, to help make the impacts more relatable to readers and to improve the linkage with the national scope of the draft report.

Figure 1.5

Figure 1.5 is an effective graphic because it provides a national map that highlights regional distinctions. It is suggested that this type of information be illustrated more often across the draft NCA4. More specifically, this figure provides context and clear take-away messages on actions being taken to address climate change, which ties in more directly to the focus of this report than other figures in the draft Chapter 1. The figure could be made more effective by magnifying the size of the orange dots, even if it creates some overlap. In the caption it would also be useful to list the total number of orange dots in the figure and to indicate what gray shading represents. Updating the figure to provide information (or a clearer distinction) between states that have renewable energy portfolio standards (legally binding) and those that have renewable energy goals (not generally legally binding) would also be beneficial. These suggested edits also apply to the inclusion of this figure in Chapter 29 as Figure 29.1.

CHAPTER 2: OUR CHANGING CLIMATE

Summary

Chapter 2 provides a high-level summary of the CSSR as background on observed and projected changes in the climate system and serves as technical input to other chapters in the draft NCA4 report. Key messages are used to summarize the key findings of the fifteen chapters of the CSSR. These key messages are supplemented by discussions in seven boxes covering topics including natural variability, climate change indicators, greenhouse gas emissions targets, extreme events, the 2017 hurricane season, and climate models. Overall, the key messages in Chapter 2 represent the key findings conveyed in the CSSR very well and provide a description of some of the major elements of global and regional climate change, with an emphasis on how those changes might affect the United States. There are no major concerns about this chapter, but the authors may consider some areas for improvement.

Review Comments Related to the Statement of Task

Comment on Key Messages

The authors of Chapter 2 have arranged the key messages in an order that differs somewhat from the chapter order in the CSSR. However, the order makes sense: the chapter starts with global or large-scale changes (e.g., global mean changes, ocean changes, and sea level rise), followed by national changes (e.g., temperature, precipitation, and floods/droughts), followed by larger-scale changes (e.g., Arctic, atmospheric circulation, and ocean circulation) that influence the United States. Each key message is described and followed by more detailed discussions, which are supported by references to key literature. The contents of the seven boxes is well chosen to supplement the key messages and address commonly raised questions about climate change.

> **Key Message 6:** Annual precipitation has increased across most of the northern and eastern United States and decreased across much of the southern and western United States; these regional trends are expected to continue over the coming century. Observed increases in the frequency and intensity of heavy precipitation events in most parts of the United States are projected to continue. Surface soil moisture over most of the United States is likely to decrease, accompanied by large declines in snowpack in the western United States and shifts to more winter precipitation falling as rain rather than snow in many parts of the central and eastern United States.

Key Message 6 mentions "shifts to more winter precipitation falling as rain rather than snow in many parts of the central and eastern United States." Such shifts are also expected in the western U.S., so it is not clear why only the central and eastern United States are highlighted. The Committee suggests revising the messages to provide a more complete explanation of where the changes are expected.

The discussion of flood and drought changes provided for Key Message 6 is somewhat confusing because it lacks discussion of the many natural factors and human activities that influence flood and drought. Precipitation, temperature, and evaporation changes are only a few of the many natural factors that contribute to these events. In addition to emissions of heat-trapping greenhouse gases caused by human activities, land use and land cover change, as well as water use and management, affect flood and drought. As such, it should not be surprising that flood and drought trends have not been consistently detected.

With an emphasis on extremes including floods and droughts in Key Message 6, it may be useful to call upon past climatic anomalies in the instrumental record (e.g., the Dust Bowl, hurricanes in the eastern seaboard, and the 1983 El Niño), and during the pre-instrumental period as inferred from proxy measures (e.g., medieval drought) in the supporting text.

Comments on Graphics

For Figure 2.2 in the draft NCA4, the y-axis title indicates the graphic is showing temperature change relative to a 1986-2015 baseline while the caption lists the baseline as 1986-2005, suggesting that one of these years is a typo and should be corrected. This also applies to the use of this graphic as Figure 1.3 in the draft report.

Comments on Literature Cited

The chapter makes good use of peer-reviewed literature and represents an accurate and well-written summary of the CSSR and the current understanding of global and regional climate change.

Comments on Traceable Accounts

The traceable accounts adequately and accurately explain how the key messages were derived.

Other Recommended Improvements

In the context of this chapter, the term "human activities" seems to refer mainly to those activities that emit greenhouse gases, but it is well known that human activities that emit aerosols change the land surface properties and alter land surface biophysical and hydrological states, which have important effects on temperature, precipitation, flooding, drought, etc. It would be useful to clarify the scope of human activities in this chapter, as the narrow definition of human activities presented seems to perpetuate in other chapters of the report.

This chapter could use more explanation of the mechanisms or reasons for the observed and/or projected changes in order to demonstrate the strong scientific underpinning of global and regional climate change. For example, an explanation of the Clausius-Clapyeron relationship would be useful for the readers to relate precipitation changes with temperature changes. Similarly, an explanation of the potential intensity (the theoretical limit of the maximum intensity that can be achieved [Emanuel, 1999]) would be useful for the readers to understand how tropical cyclone intensity changes can be related to sea surface temperature warming and why there is more uncertainty in projecting tropical cyclone frequency changes than intensity changes. Similarly, citing the conceptual knowledge gleaned about global temperature, sea level, and biosphere response from paleoclimate records and longer time horizons of change would help the readers understand the various changes. For example, Clark et al. (2016) provides a different conceptual view of sea level rise following warming events, which is a useful discussion to cite.

This chapter could also include more discussion of the oceans related to carbon dioxide (acidification), warming (ocean heat content), and changes in large-scale circulation and precipitation (e.g., sea surface warming patterns have an important influence on precipitation changes) that have impacts that are appropriate to discuss in the draft NCA4.

Box 2.1 provides a useful discussion of natural variability as a source of uncertainty in understanding past changes and projecting future changes. However, it would be useful to also discuss uncertainty more broadly, and perhaps call it out in a separate box because climate projection uncertainty also includes model and scenario uncertainty. Alternatively, the draft NCA4 Chapter 1 could be a place to expand this discussion and reference it prominently in Box 2.1.

Box 2.7 should acknowledge recent advances in dynamical downscaling at sub-10 km resolution that more explicitly resolves convection. A good example to reference is Prein et al. (2016, 2017), which generates decade-long regional projections of precipitation changes in the

United States. A relatively large body of research is now available, supporting the transformative capability to model regional climate using convection-permitting modeling. Also, a mention that there are important variables or phenomena for which climate changes are still not well understood or downscaled using dynamical or statistical methods (e.g., wind, cloudiness, sea breeze, Santa Ana winds) would be useful for the readers in order to appreciate why certain local/regional phenomena are not discussed.

Chapter 2 could include some discussion of changes that are more controversial but capture the interest of the public. For example, warming is expected to increase the frequency of heat waves. However, cold extremes may still occur because of possible changes in blocking and extreme snowfall may not decrease by as much as expected based on temperature alone (e.g., O'Gorman, 2014). This information could then be drawn upon in other relevant chapters of the draft NCA4.

Long term global and regional observations of the coupled climate system play a crucial role in tracking and understanding changes. This may warrant a separate discussion or box, including salient examples.

An additional box summarizing the key advances in understanding, modeling, and observations of climate change since NCA3 would be useful for readers who want a cursory look at what's new in NCA4 relative to NCA3. The CSSR report has a nice summary of "Advances Since NCA3" in Box 1.2 of that report (USGCRP, 2017) that the Chapter 2 authors could draw on.

CHAPTER 3: WATER

Summary

Chapter 3 is generally well written and succinctly covers an impressive amount of material on the complex topic of water. The "Regional Roll-Up" section overviews are a good complement to the general national-level information. The chapter is particularly strong on the topic of climate change implications for existing and planned infrastructure, including some strong examples of proactive management actions. The chapter generally makes appropriate reference to the rich literature on these topics. In terms of areas of improvement, the information on direct biophysical impacts to the water cycle is generally underdeveloped compared to the subsequent discussions of mitigation and management. Overall, this chapter would benefit from discussions on how the state of the science (and/or data or management practices) either has or has not evolved since the NCA3. This addition should highlight recent advances in understanding climate change impacts on hydrology, floods, and drought.

In discussing human influence, there is a need to clarify the role of human activities (e.g., emissions of greenhouse gases) that lead to warming (and subsequent changes in the water cycle) versus human activities such as irrigation and water management that have direct impacts on hydrology. This distinction is relevant to attributing historical hydrological changes, as well as projecting future hydrological changes. For example, warming-induced drought and stream temperature rise could be alleviated by water management actions that increase flow (e.g., Wan et al., 2017).

The chapter could also greatly benefit from more direct and explicit discussions connecting to the climate change projections described in Chapter 2, "Our Changing Climate," of the draft NCA4 report.

Comments Related to the Statement of Task

Comments on Key Messages

In general, the key messages in this chapter are not all well linked to Chapter 2 of the draft NCA4. For example, groundwater depletion and water use are noted prominently in Chapter 3, but were not mentioned in Chapter 2. Conversely, Chapter 2 has some good discussions of flood and drought, while Chapter 3 only briefly mentions the changes in snowpack and rain-to-snow ratio and speculates how increased extreme precipitation may result in more severe flooding. An explicit reference to the draft NCA4 Chapter 2 in the Water Chapter about how climate affects hydrology, floods, and drought would be useful to improve the connection and provide a more complete picture.

> **Key Message 1:** Significant changes in water quantity and quality are evident across the country, presenting a risk to coupled human and natural systems and related ecosystem services. Rising temperatures are reducing snow-to-rain ratios, leading to significant differences between the timing of water supply and demand. Groundwater depletion is exacerbating drought risk. Surface water quality is declining as water temperature increases, and more frequent high-intensity rainfall events mobilize pollutants such as sediments and nutrients.

Key Message 1 highlights the changes in water quantity and quality, with a focus on robust changes that are well supported by numerous studies. These include reduced snowpack and increased rain-to-snow ratio with warmer temperature; increased human water use due to warmer temperature and other factors; groundwater depletion (due to agriculture); warmer stream temperature; extreme precipitation mobilizing more transport of sediment and nutrients by rivers; sea level rise and salt water intrusion; and wildfires and other factors that affect water quality. These are all important findings that are relevant to the observed and projected changes in many regions, so this key message is very well connected with many regional chapters. However, the text supporting Key Message 1 could be improved by rebalancing considerations of temperature and rainfall changes. Temperature warming is currently the disproportionate focus in the discussion (e.g., changes in snowpack, rain-to-snow ratio, stream temperature), whereas impacts of projected changes to extreme precipitation events (e.g., Wehner et al., 2013; Westra et al., 2013; Feng et al., 2016) and dry spells (e.g., Peterson et al., 2013) are less developed. A brief discussion about the impacts of saltwater intrusion on drinking water treatment could also be added, as well as information on how water utilities relying on ground and surface waters may be affected (e.g., Kolb et al., 2017).

The paragraph about groundwater depletion (page 133, line 27, to page 134, line 5) needs more care to explain the main point: that reduced groundwater availability exacerbates drought risk. As currently written, some readers may come away with the impression that the recent groundwater depletion trends discussed here are a direct outcome of past climate changes, which they, for the most part, are not. Rather, they are mostly associated with agricultural intensification and some other sectoral usages. When considering climate change impacts on future groundwater storage, it is important to consider both the demand (pumping) and the supply (recharge); there is substantial literature that considers both sides of the water balance that could be referenced. Regarding importance of groundwater sustenance, the Chapter 3 authors should consider adding a comment about groundwater-dependent ecosystems.

> **Key Message 2:** Aging water infrastructure compounds the climate risk faced by society. Extreme precipitation events are projected to increase in a warming climate and may lead to more severe floods and greater risk of infrastructure failure in some regions. Infrastructure design, operation, financing principles, and regulatory standards typically do not account for a changing climate, presenting a risk to existing infrastructure systems. Current risk assessment methods also do not typically consider the impact of compound extremes (co-occurrence of multiple events) and the risk of cascading infrastructure failure.

Key Message 2 is well supported by the cited literature and makes good use of examples. A reference to the draft NCA4 Chapter 17, "Sectoral Interdependencies, Multiple Stressors, and Complex Systems," would be useful, as water is a key component of the water-energy nexus that requires cross sectoral considerations. Additionally, a few references are made to the representation of extreme hydrologic events in the paleo-hydrology literature, including in the text supporting this key message in relation to infrastructure design standards. The Chapter 3 authors should clarify whether they are recommending incorporation of paleo-environmental information into water infrastructure design practices.

> **Key Message 3:** Water management strategies designed in view of an evolving future that we can only partially anticipate will help prepare the nation for the water and climate risks of the future. Current water management and planning principles typically do not incorporate the ability to address risk that changes over time. There are positive examples of promising directions to manage climate vulnerabilities, while the gap between research and implementation, especially in view of regulatory and institutional constraints, remains a challenge.

Key Message 3 focuses on water management in a changing future and highlights the scientific challenge as well as institutional, political, and legal challenges and the need to manage vulnerabilities for a wide range of uncertain conditions. The text supporting this key message provides robust reference to the literature. Figure 3.3 is an effective illustration of the issues surrounding the balance of water supply and demand and the uncertainty. As noted for Key Message 2, reference to the draft NCA4 Chapter 17 would be useful here, as water must be managed in the context of not only providing water resources but also considering energy, agricultural, and other sectors.

Uncertainty should be included in Key Message 3 and not just in the supporting text that follows the message. The discussion about uncertainty—using the Great Lakes and other places as examples—is strong. This addition could better convey that uncertainty in projecting future changes is a major obstacle to addressing risks that change with time.

Comments on Literature Cited

Generally, the appropriate literature is well cited in Chapter 3, although some specific references to add are noted throughout this chapter review.

Comments on Graphics

Figure 3.2 is misleading because it suggests uniform depletion across the vast Ogallala region. In reality, the northern part of the aquifer system has experienced relatively stable groundwater levels, whereas the southern portion has undergone extensive overdraft and depletion. It is a common point of confusion and the text should be updated to clarify this regional distinction.

Comments on Traceable Accounts

The Chapter 3 authors assert that formal attribution of flood and drought to human-induced climate change has not been established. This statement is somewhat at odds with the authors' assignment of "medium confidence" to more severe floods in the future. This nuance requires more discussion to avoid confusion. For example, formal attribution of flood and drought to human-induced climate change may not be possible without considering human activities such as irrigation, groundwater use, and water management. In this chapter as well as others, the term "human activities" is often used to attribute or project future changes, but only those changes related to the emission of greenhouse gases such as fossil energy use and land cover and land use change. It is important to clearly distinguish human activities related to emission of greenhouse gases from human activities related to irrigation and water management. The latter have direct impacts on floods and drought (both agricultural and hydrological). For example, water management practices that regulate streamflow for flood protection and low flows could reduce the detection of increasing floods/droughts. This can have broader implications if the reduced detection is combined with aging infrastructure such that protection against flood and drought in the future is lessened, increasing risks.

Other Recommended Changes

The "Chapter Development" section provided in the traceable accounts indicates that Chapter 3 was developed to place more emphasis on vulnerability, risk, and management than in the NCA3 report, since the NCA3 focused largely on climate change impacts on hydrology, flood, and drought. While this is an effective approach to keep the chapter succinct, it is still important to summarize key information in order to provide necessary context for the reader. A brief discussion regarding climate change impacts on hydrology, floods, and drought, as well as highlighting the advances made since NCA3 would strengthen the chapter. For example, since NCA3, many new aridity studies have been published, suggesting increased aridity due to warming (Sherwood and Fu, 2014) and that care should be taken in using offline approaches for projecting aridity changes (Milly and Dunne, 2016). Some studies have investigated the changes in flood (Das et al., 2013; Mallakpour and Villarini, 2015), flood seasonality (Ye et al., 2016), and drought (Cook et al., 2015). Several studies have investigated the relative impacts of climate change and human water use (e.g., water management) and suggested that because water management could mitigate drought changes in the future, projections of future drought characteristics should consider both climate change and direct human influences on the water cycle (Wanders and Wada 2015; Wan et al., 2017). It is important to convey the importance of human impacts through water use and water management when attributing past changes in flood and drought and when projecting future changes.

There are many major water management "hotspots" in the United States that are anticipated to be climate-sensitive that are not mentioned or well developed in Chapter 3. It is beyond the scope of this chapter to discuss them all, but the Chapter 3 authors should consider some additional comments or case studies about other problem areas, as space allows. Such concrete case examples tend to be very useful for illustrating climate change impacts to the reader (e.g., California Central Valley groundwater depletion, Floridian Aquifer salinization, Chesapeake Bay ecosystems).

Glaciers should be discussed in a national topic chapter because impacts are evident and there is high confidence in glacial changes. The Water Chapter seems like a reasonable chapter in which to include brief discussion of glacial melt, with cross-reference to the Alaska regional chapter as appropriate.

In this chapter, as noted for the draft NCA4 as a whole, increased consistency in the discussion of similar topics and cross-referencing between relevant sections of this chapter and both regional and other topic chapters' key points regarding water is needed.

CHAPTER 4: ENERGY SUPPLY, DELIVERY, AND DEMAND

Summary

Chapter 4 provides a strong overview of climate change implications on the nation's energy system and the actions that are underway to protect energy security and promote energy sector resilience. The text details the challenges that industry and government (at all levels) face as they attempt to ensure that energy, which underlies most U.S. economic activity, is not significantly disrupted by extreme weather and climate change. It also discusses actions necessary to escalate the pace, scale, and scope of efforts to ensure the safe and reliable provision of energy now and in the future.

The Committee has some suggestions for improvements. The chapter should include more discussion about distributed generation, including that which is relevant to renewable energy. Specifically, the section about energy sector transformations should address how renewable energy sources could be impacted by climate change and extreme weather, and how these impacts could affect or protect other associated systems. Also, because of the nature of energy ownership, the importance of private-sector actions, public-private partnerships, and new financial models should be further highlighted, potentially as part of a key message. Chapter 4 includes information about interdependencies and should cross-reference Chapter 17 of the draft NCA4, "Sectoral Interdependencies, Multiple Stressors, and Complex Systems."

Comments Related to the Statement of Task

The chapter's coverage of the energy sector is appropriate, but further consideration of the interconnected nature of the energy sector as more of an energy system or process spanning from sources, to generation, to consumption, and any feedback impacts in the context of system dynamics might be insightful. For example, the impact of Hurricane Katrina on fuel production and refining could be expanded. Linkages with other climate change impacts could also be explored and relevant chapters in the draft NCA4 cross-referenced. Interconnected topics that may impact the energy sector include climate impacts on water systems, increased salinity and changes in wildfires. Cross-referencing to relevant chapters discussing these issues should be considered.

Comments on Key Messages

The key messages are generally clear, consistent, and at an appropriate technical level for the intended audiences. Key Messages 1 and 2 affirm current understanding of: the critical nature of energy supply and delivery systems and their interconnectedness and interdependencies with other vital systems; the changing nature of energy supply and delivery systems and roles of energy technologies, markets, and policies in affecting the systems' vulnerabilities to climate change and extreme weather; critical systems that are becoming more interconnected. However, an examination of the subject matter as an energy process might be useful.

> **Key Message 3:** Actions are being taken to enhance energy security, reliability, and resilience with respect to the effects of climate change and extreme weather. This progress occurs through improved data collection, modeling, and analysis to support resilience planning, and the deployment of new, innovative energy technologies for hardening energy assets against extreme weather hazards. Although barriers remain, opportunities exist to enhance energy systems resilience.

Key Message 3 should also highlight that the listed actions are being taken by both the public and private sector as well as public-private partnerships. In addition, the key message should directly call out the use of risk methods to inform policy and decision-making practices for achieving energy and economic efficiencies.

Comments on Figures

Overall the graphics are clear, internally consistent, and communicated well for the intended audiences. Figure 4.1 is interesting and informative and Figure 4.2 illustrates critical infrastructure interdependencies and is an appropriate choice to include in the chapter.

Figure 4.3 is interesting, although it might be too specialized for a non-technical reader. A figure that shows more generally the various impacts of the energy system or the energy process and associated impacts from exploration and generation to distribution and consumption would be useful for the chapter.

Comments on Literature Cited

The chapter accurately reflects the peer-reviewed scientific literature with a particular focus on literature since the NCA3. No critical content areas were left out from the chapter.

However, as recommended earlier, more discussion of the underlying process may be an insightful addition for readers.

Comments on Traceable Accounts

The traceable accounts underlying the key messages for Chapter 4 provide consistent, transparent, and credible documentation of the foundational literature. The key messages accurately reflect supporting evidence and are communicated effectively.

Other Recommended Changes

Because energy supplies are typically privately owned and operated, private sector actions should be discussed in more detail and the importance of public-private partnerships should be highlighted. Potential changes to risk management and adaptation also require new planning and design philosophies that rely on adaptive-type methods. One framework that could be mentioned is "Real Options" (Zhao et al., 2004; Chiara et al., 2007), which provides an economic assessment framework for policy and decision-making.

The energy system includes primary fuel supply, power generation, and transmission and distribution systems. Throughout the energy system, infrastructure is aging and frequently exceeding design lifespans. The chapter could be improved by making reference to related issues and system exacerbations from a changing climate in order to provide a more complete understanding of how climate change impacts interact with other existing challenges in maintaining the energy system. In addition, while the chapter appropriately discusses electricity sector vulnerabilities, it should also mention that increased demand for distributed sources (including renewables electricity) may make management of these multiple facilities more challenging. Renewable energy systems also have their own challenges, including, but not limited to, assuring adequate biomass (especially in times of drought) and system stability for wind and solar energy systems during extreme weather events and changes in regional climate.

CHAPTER 5: LAND COVER AND LAND USE CHANGE

Summary

Chapter 5 does a nice job of linking literature on land cover effects on climate and climate impacts on land cover and land use. This is done especially well through focusing on impacts on disturbance regimes, species distributions, and land use suitabilities. The chapter also provides an overview of how changes in land cover can have both mitigation and adaptation benefits. It is appropriately detailed in its review of both implications on climate change and from climate change.

The key messages cover important interactions between land cover and land use and the climate, but they could be articulated to better reflect the main takeaways. The NCA3 chapter on this topic explicitly called out mitigation and adaptation opportunities associated with land use. These topics are addressed throughout the chapter in the draft NCA4, presumably in the interest of reducing chapter length. Because land cover and land use approaches to adaptation and mitigation are not consistently handled within other national topic and regional chapters, this represents a missed opportunity. While land management changes have been shown to help with adaptation and mitigation, such as in the agricultural sector, some land use change will likely need to occur as productivities are affected by climate and soil restoration activities are prioritized. This point is not clearly addressed anywhere in the draft report and should at least be articulated well in Chapter 5. Chapter 5 also includes a description of the state of the sector along with land use projections. The land use projections, however, are poorly described and are not well integrated into this chapter or the other relevant chapters in the draft NCA4. In addition, linkages from the draft NCA4 Chapter 5 to other chapters, and linkages among the various sections within Chapter 5 (including confidence statements) are not consistent.

Comments Related to the Statement of Task

Comments on Key Messages

> **Key Message 1:** Changes in land cover, which may be driven by societal choices concerning land use, continue to impact local- to global-scale weather and climate by altering the flow of energy and water between ecosystems and the atmosphere, with important feedback effects on the climate system.

Key Message 1 lacks clarity in two respects. First, the message should be clearer about the distinction between effects on local climate (e.g., through water and energy balance impacts, as highlighted in the message) and effects on greenhouse gas emissions, which are also raised in the text, but not in the message. Second, because feedbacks are by definition bidirectional, the message should better articulate what the feedbacks are "between the climate and land systems" rather than "on the climate system."

The supporting text for Key Message 1 includes evidence for both impacts on water and on energy balances. These impacts modify climate locally and affect changes in greenhouse gas emissions, which modify climate globally. The key message seems to focus on the former, so the supporting text should as well.

> **Key Message 2:** The composition of the natural and human landscapes, and how society uses the land, affects the ability of the Nation's ecosystems to provide essential goods and services. However, climate change is expected to directly and indirectly impact land use and cover by altering disturbance patterns, species distributions, and suitability of land uses.

Key Message 2 could be clearer if it started with a simple statement that "climate change affects land use and ecosystems." While this point is in the message, it comes across as secondary and should be more prominent.

Comments on Graphics

Figure 5.1 is busy and difficult to read. The figure could be simplified by including only one data product for land cover on the figure or by selecting a different graphic. The two land-cover products are not adequately different to justify including both. Given the available time series, it is suggested that the results of the National Land Cover Database product be presented, and that full reference(s) to all data sources in the figure be provided, which are not included in the draft chapter.

Figures 5.2 and 5.3 are too small and busy to easily decipher. Their size could be increased to improve graphic clarity.

Comments on Cited Literature

Generally, the appropriate literature is cited in Chapter 5, but a few additions are recommended. For albedo implications, Jones and Hawkes (2015) should be referenced. Also, discussion of forecasts of changes in irrigated agriculture should probably consider recent trends and regional differences (see Brown and Perves, 2014).

Comments on Traceable Accounts

Findings are generally consistent and transparent, though some inconsistencies are noted in the line comments for clarification (see Appendix B of this review report).

Other Recommended Improvements

A better description and definition of land cover scenarios in the section on future projections is needed. Sleeter et al. (2017) seem to be doing straight-line projections of trends for a "business-as-usual" (BAU) scenario. BAU might imply some (absent) process model and the scenario should perhaps be renamed as "current trends projection" and be better described. None of the projections are referenced elsewhere in the chapter or used to evaluate possible future effects on climate impacts (e.g., in the coastal zone) or on carbon storage.

Chapter 5 could be enhanced by discussion of the potential positive implications of climate for land-use (agriculture and ecosystems), e.g., increased growing seasons (the benefits of which are reduced by increased populations of pests).

It would be helpful if the topics discussed could be better cross-referenced with other chapters. For instance, increased treatment of the implications of fire (draft NCA4 Chapter 6, "Forests") and mention of land cover and land use changes that are occurring in the Arctic region (draft NCA4 Chapter 26, "Alaska") should be included.

Key messages from this chapter are not specifically reflected in the draft NCA4's overarching report findings. Land cover and land use change are processes that interact with climate change, so it is appropriate that they are not included as report findings, but these interactions need to be acknowledged (in some cases better than they are currently) throughout the report.

CHAPTER 6: FORESTS

Summary

Chapter 6 is well written and clear about the impacts of climate change in recent decades and what some future adaptation options for forests may be. It describes the direct consequences of climate change (regeneration growth and productivity, mortality, and range shifts) as well as the indirect effects (wildfire, pathogens, insects, and carbon storage) in a clear fashion. The chapter considers projections of warmer temperatures, reductions in snowpack, increased drought, and increased extreme climate events in its analysis. It also integrates, evaluates, and interprets findings from recent assessments of forest health and drought resiliency. In these respects, the chapter makes general use of the findings of the CSSR and the draft NCA4 Chapter 2. The traceable accounts show that these climate-induced changes are representative of the literature.

A general concern about the chapter is the need for a statement on forest vulnerability that recognizes that forest types, disturbance regimes, and management objectives are highly variable across the United States and that no single management response or adaptation strategy will achieve resilience in all forest types. Forest differences are referred to in the chapter, but the majority of examples emphasize pine-dominated and commodity forests, with little consideration of issues relevant to urban, deciduous, mesic evergreen, tropical, and high-elevation forests. The Committee recommends an opening paragraph describing the diversity of U.S. forest types, disturbance regimes, ownership (public versus private), and uses (commodity, recreation, urban, wilderness), as well as a statement that explains which forests or regions are the focus of the chapter (e.g., there is currently no discussion of forests in Alaska, Hawai'i, or U.S. territories).

The chapter promotes fuel treatments as an adaptation measure to reduce wildfire risk, but again, this strategy is only appropriate for some, not all, forest types. Several studies suggest that fuel treatments have been unsuccessful or are economically unfeasible at a broad scale in many U.S. forests (e.g., Cochrane et al., 2012; Fulé et al., 2012; Butler et al., 2013; Barnett et al., 2016; Bradley et al., 2016; Kalies et al., 2016). Additionally, fuel treatments are unlikely to succeed in the future in light of more extreme fire weather. The success of fuel reduction efforts should be discussed in conjunction with Key Message 1 (greater disturbance with increased frequency of extreme weather). The line comments for this chapter (in Appendix B of this review report) note places in the draft chapter where more geographic or ecological specificity would be helpful, as well as parts of the text where technical terms should be replaced with nonspecialized language to improve readability for the intended audience.

Comments Related to the Statement of Task

Comments on Key Messages

Generally, the key messages for Chapter 6 appropriately reflect the primary issues related to climate change for U.S. forests, although some edits are recommended.

Key Message 1: It is highly likely that more frequent extreme weather events will increase the frequency and magnitude of severe ecological disturbances, driving rapid (months to years) and often persistent changes in forest structure and function across large landscapes. It is also likely that other changes, resulting from gradual climate change and less severe disturbances, will alter forest productivity, health, and the distribution and abundance of species at longer time scales (decades to centuries).

The chapter and supporting text for Key Message 1 does a good job of describing recent changes in disturbance regimes (more wildfires, insect outbreaks, etc.). The mention of more gradual changes in forest productivity, health, and distribution and abundance in the coming decades to centuries makes sense, but the evidence is less clear. See Appendix B of this review report for additional suggested citations to support this key message.

Key Message 2: It is highly likely that climate change will mostly decrease the ability of forest ecosystems to provide ecosystem services to society. Tree growth and carbon storage are expected to decrease in most locations as a result of higher temperature, more frequent drought, and increased disturbances. The onset and magnitude of climate change effects on water resources in forest ecosystems will vary but are already occurring in some regions.

By citing only tree growth and carbon storage, Key Message 2 does not address other important ecosystem services provided by forests, including protection of native biodiversity (including endangered species), provision of important resources for indigenous communities, places for recreation, and maintenance of clean water. The key message should be revised to include a more representative, concise list of services.

Key Message 3: Forest management activities that increase the resilience of U.S. forests to climate change are being implemented, with a broad range of adaptation options for different resources, including applications in planning. The future pace of adaptation will depend on how effectively social, organizational, and economic conditions support implementation.

Key Message 3 is somewhat unclear as written. In particular, the phrase "broad range of adaptation options for different resources, including applications in planning" is ambiguous. One suggested revision to address this is: "Forest management activities can increase the resilience of U.S. forests to climate change, and a broad range of management and adaptation options are currently being implemented to protect, maintain and enhance different forest resources. The future pace of adaptation will depend on how effectively…." Key Message 3 should also mention "responses" in addition to resilience, to present a more fully developed conceptual model for thinking about long-term climate impacts in forest ecosystems. Resilience also implies that forests will return to their original state after a short-term disturbance. Large climatic changes however, may cause forests to transition to new states, which will influence adaptation strategies (Millar et al., 2007).

Key Message 3 also does not address risk. It seems important to mention the notion of learning to live with fire in the future in order to better align with Key Message 1. There are several management efforts underway, for example, to reduce wildfire risk for communities in the wildland-urban interface. The examples offered to support Key Message 3 are largely federal programs; it would strengthen the text to mention some of the successful public-private partnerships and state and community efforts that seek to maintain healthy forests and reduce risk. The Committee thinks there is also an overemphasis on fuel treatments and thinning as ways of increasing forest resilience. It should be stated that these management strategies are appropriate for some forests, when applied in a targeted fashion to reduce fuel loads, especially near structures. These strategies will also be most successful in reducing wildfire risk under non-extreme conditions. Maintaining fuel treatments is not practical on a broad scale, especially given the extreme climate events discussed in Key Message 1. See Appendix B for additional suggested citations to support this key message.

Comments on Graphics

The figures are internally consistent and are clear, but a few edits are suggested.

The caption for Figure 6.4 should explain exactly what is shown. Fire starts or beetle outbreak? Area burned? Insect-related tree mortality?

Figure 6.5 should clearly indicate that it is developed for the Pacific Northwest and the examples are region-specific. The beaver photo (page 237 of the draft NCA4) should indicate the location and also the management program that is implementing reintroduction of beaver communities.

Comments on Literature Cited

The chapter provides many recent citations from the peer-reviewed scientific literature that are very effective. It would be helpful if this new information was used to more clearly identify advancements in research since the NCA3. The Chapter 6 authors could also consider citing additional recent publications, particularly those relevant to the western U.S. forests, wildfires, and risk.

The authors may consider referencing paleoecological literature that demonstrates forest sensitivity to changes in temperature and precipitation in the past, as well as wildfire disturbance. Recent publications related to fire trends and legacy effects, rapid forest change and disturbance synergies (Key Message 1), long-term forest change (Key Message 1), and fire adaptation and risk assessment (Key Message 3) are provided in Appendix B of this review report, in addition to some paleoecological studies.

Comments on Traceable Accounts

Overall, the findings are documented in a transparent and credible way and they reflect supporting evidence. As detailed in the line comments (Appendix B of this review report), some specific information and qualification are needed to accurately portray study-specific evidence and parts of the draft chapter where additional citations might strengthen the information, especially for western U.S. forests, are noted. The assessment of confidence is communicated effectively.

Other Recommended Improvements

In general, some terminology in the chapter should be clarified. Basic terms like "large-scale" versus "small-scale," "healthy forests," "fire and other regimes," "historical range of variability," and "climate-smart forest management" should be defined for a non-specialist audience. Similarly, examples that reference published studies should include information about the geographic scope (region, forest type, watershed) and the time span of those investigations. Phrases like "in some forests…," "in some regions…," or "recently" are too non-specific.

The formatting for citations in the chapter is inconsistent. Single authored citations for multi-authored papers are sometimes shown and some references are missing. In a few places, the citation does not accurately support the statement or represents a very limited finding.

Inclusion of Chapter 6 Materials in Chapter 1, "Overview"

Negative impacts on timber prices and economic well-being of forest landowners (page 36, lines 16-20) are not discussed specifically in Chapter 6, except by reference to a single paper (page 229, line 35). It would be useful to have more supporting information on these negative impacts in Chapter 6 in order to justify their inclusion in the draft NCA4 Chapter 1. Similarly, the effects of ground-level air pollutants (page 44, line 13), are not discussed in Chapter 6 except by a single reference (page 230, line 24) and the negating influence of ground-level ozone is not well explained—what is the source and what are the consequences?

In general, the Committee was somewhat disappointed that Chapter 1 does not include more reference to forests (relative to agriculture or infrastructure, for example). This omission might be in part because of some of the shortcomings in Chapter 6 described in this review. For example, if the forest chapter had described how climate impacts on forests played out differently in different regions or how adaptation measures (e.g., fuel treatments) must differ among regions, perhaps these points would have appeared in Chapter 1. The draft Chapter 6 contains important general findings that probably characterize many sectors.

CHAPTER 7: ECOSYSTEMS, ECOSYSTEM SERVICES, AND BIODIVERSITY

Summary

The draft Chapter 7 focuses on a subset of topics related to ecosystems, ecosystem services, and biodiversity, with a particularly heavy emphasis on species responses. The key messages in the draft chapter are overly generalized. The chapter would benefit considerably from expanded treatment of ecosystem services and discussion of climate mitigation strategies. Chapter 7 should also be linked more closely to related topics in the draft NCA4. Many strong examples of ecosystem services presented elsewhere in the draft report could be discussed and are noted in this chapter review.

Comments Related to the Statement of Task

Comments on Key Messages

Overall, the key messages reflect understanding about observed and projected impacts, but the evidence is presented with too many generalities. Providing some specific examples would help to support the information in the chapter's key messages.

> **Key Message 1:** The resources and ecosystem services that people depend on for livelihoods, protection, and well-being are increasingly at risk from the impacts of climate change. Climate change has already had observable impacts on biodiversity and ecosystems throughout the United States. Both marine and terrestrial species are responding to climate change by expressing different traits, altering behaviors, shifting ranges, and changing the timing of biological events. Climate change may outpace the rate at which species can adapt. Projections suggest many shifts could substantially alter change species interactions, and create mismatches in resources, and reconfigure ecosystems into novel assemblages with uncertain consequences for ecosystem function and services.

For Key Message 1, the links to climate change are made in a general way, with few examples that will stick with a reader. This shortcoming could be improved by including in the message the key elements of the "ecosystem function and services" that are likely to be affected. This key message is also quite long and may be more effective if split into two messages with one focused on systems, communities, and ecosystem responses, and a second focused on ecosystem services.

> **Key Message 2:** Natural resource management will increasingly require planning for an uncertain future. Adaptation strategies that are flexible and coordinated at landscape and large marine ecosystem scales have rapidly progressed and their implementation is continually being refined to address emerging impacts of climate change and how they are compounding with other stressors on our valued resources.

For Key Message 2, some specific examples (e.g., key species, habitat types, and ecosystem services needing extra attention or showing promising responses to management) would help support the information, which is otherwise vague. Successful management strategies and programs provided as support for Key Message 2 are simply a list of mostly federal programs with no detail. A table providing expanded information would be helpful, and it would be good to list other efforts by states, public-private partnerships, and nongovernment organizations.

It is also unclear why mitigation strategies are not mentioned in Key Message 2. Changes in biodiversity and ecosystem services can have a large impact on carbon sequestration and storage, as is cited in other chapter key messages. Furthermore, it is surprising that climate resilience is not included, given that climate refugia and habitat connectivity are mentioned under adaptation strategies, mostly highlighting invasive species threats. Resilience is referred to throughout the text, but never defined or illustrated.

Comments on Comments on Graphics

The chapter is under-illustrated and Figure 7.1, the only figure, provides anecdotal information about biotic and abiotic changes related to climate change around the country. The examples provided should be discussed in the text with citations to the original publications.

The Committee suggests that the Chapter 7 authors review and consider including updated versions of some figure(s) including those in Staudinger et al. (2012), which served as a technical input to the NCA3. Figures to review could include 2.1, 3.2, 3.7, and 4.2.

Using boxes or other examples to make the general concepts presented in the chapter more understandable and accessible would go far to improve this chapter.

Comments on Literature Cited

The effects of temperature, lengthened growing season, drought, extreme events, and carbon dioxide enrichment have clear influences on ecosystems, yet there is little direct reference to relevant literature, or to the draft NCA4 Chapter 2, "Our Changing Planet." It is recommended that these relationships be highlighted in the chapter by providing examples of management strategies that target specific ecosystems and ecosystem services (and appropriate citations). More detailed recommendations are provided later in the section.

References for other ecosystem services, in addition to pollination services, should be included to bolster general statements. Also, as stated in comments for Key Message 2, there is no mention of the extensive literature on climate resilience. Some references are also missing and a few are miscited, warranting careful review by the Chapter 7 authors. The paleo references are also over-attributed.

There are too many generalities and references to meta-analyses, rather than the primary literature. Specific examples carry significant weight in convincing stakeholders about the particular topics that are the focus of this chapter. A number of examples that the Chapter 7 NCA4 authors could cite are listed in the "Other Recommended Changes" section of this chapter review and in the line comments (see Appendix B of this review report).

Comments on Traceable Accounts

In general, the traceable accounts need better documentation to support the key messages. There is little specific information about how climate change has impacted species or ecosystems compared to other non-climate drivers of change. Key Message 1 reads like a primer of potential future changes. Key Message 2 is very general and focuses on federal agency response without considering the work of state agencies and non-governmental organizations. Some examples are needed.

Also, the description of evidence for Key Message 2 (pages 273-274) does not align with the discussion in the main text of the chapter (pages 266-267) and needs some case studies, perhaps including examples of resource management actions from regional chapters discussed in a broader context. The evidence also does not describe adaptation efforts to preserve ecosystem services and biodiversity or to maintain resilience. The uncertainties listed on page 275 seem quite accurate, but rather than focus on uncertain species' responses with respect to management practices, it would be better to discuss the need to protect the provision of ecosystem services.

Other Recommended Changes

The Committee encourages the chapter authors to look at the structure of other related chapters, such as the draft NCA4 Chapter 9, "Oceans and Marine Resources," as an example of a well-developed and organized chapter.

Chapter 7 would be an ideal place to document links between climate, ecosystems, and equity/access issues for individuals and their communities. Describing loss of ecosystem services is one of the best ways to demonstrate that particular communities will be impacted by climate change. It can also highlight compelling examples of social, cultural, health, and economic disparities that result from changed ecosystem function.

The rationale for framing the chapter around the selected subset of topics (range shifts, phenology, adaptive capacity, invasive species, and emergent properties) is generally unclear. These topics are apples and oranges in terms of issues and the list omits several important topics that are well documented in the literature (see citations provided in the draft NCA4 that are quoted later in this chapter review).

Chapter 7 is missing treatment of specific climate science findings summarized in Chapter 1 of the draft NCA4. In particular:

- It is unclear why there is no treatment of climate mitigation strategies in Chapter 7, only adaptation. Biodiversity and ecosystem services can have significant impact on carbon sequestration and storage. Because the draft NCA4 covers adaptation and resiliency (Chapter 28) and mitigation (Chapter 29), it would be appropriate to discuss and link discussion of these topics in Chapter 7.
- Stressors—only invasive species are mentioned, with no mention of other stressor issues, which are summarized in the draft NCA4 Chapter 2 and in the CSSR. Other areas with a need for increased resilience include disease, increased frequency of extreme events such as fire, drought, riverine and coastal flooding, sea level rise, and ocean acidification.
- The interaction and fundamental underpinning of biodiversity and ecosystem services for topics covered in many other draft NCA4 chapters (Water, Energy, Land Cover and Land Use Change, Forests, Coastal Effects, Oceans and Marine Resources, etc.) should be highlighted in the report findings. Relevant chapters should also be cross-referenced throughout this chapter.

The draft Chapter 7 places most emphasis on species responses, especially invasive species, but does not spend much time evaluating changes in ecosystem services and biodiversity and their consequences. For example, the chapter is missing discussion and documentation of effects of climate-induced changes to biodiversity and other ecosystem services with respect to change in agriculture, coastal protection, recreation, fisheries, timber harvest, carbon sequestration and storage (climate regulation), flood control, water retention, water provision, water quality regulation, hydropower/tidal energy production, disease regulation, temperature regulation, cultural services, etc. While expanded discussion of all these topics would make the chapter unreasonably long, careful consideration of additional topics, including those that tie in well to other impacts discussed in this chapter, should be included to provide better chapter balance.

Specificity would help increase the significance of several statements: over what time span has change been noted in ranges (pages 259-260)? What are the range changes? What organisms are involved? References to examples discussed in the regional chapters would make an important connection. See Appendix B for more specific recommendations. In addition, the synergistic effects of land use change, pollution, and other human activities are not considered in assessing changes in species distributions, behaviors or interactions. Expansion of the chapter to address these suggestions would provide a more complete evaluation of the chapter's topics.

The "Regional Roll-Up" section comes across as a mish-mash of examples explained in a low level of detail. More information is needed to support the examples as evidence of climate change. For example, the statement that climate change is threatening salmonid populations in the Southwest and Northwest, without discussing species, land use, fire, and other factors, is incomplete. The reference to bear populations and reduced salmon mortality rates in Alaska is also unclear. This revision could be done as parenthetical additions to the sentences or a more logical breakdown by regions. Chapter 7 should also acknowledge the lack of regional and nation-wide modeling efforts linking changes in climate to changes in biodiversity and ecosystem services, if that is the case.

The chapter should acknowledge uncertainties and "emerging issues" and explain what is well understood, what is more speculative, and what needs more research to be fully understood.

Overall, the draft chapter is written at a level appropriate for non-specialists, although terms like "holistic ecosystem-based approaches," "harmful impacts of current and future resource management challenges," "tragedy of the commons," "seed sourcing," "novel sectors and livelihoods," and "assisted migration" should be defined.

Select Detailed Comments

A few line comments are highlighted here; see Appendix B for additional suggestions.

- Page 257 and page 259, lines 21-31 of the draft NCA4. By focusing on the "state" or "stocks" of biodiversity and ecosystems, there is a narrow focus on impacts to species and communities. Virtually no mention of which ecosystem services are likely to be affected is provided. Flows from the "stock" (biodiversity and natural ecosystem components) to affected people is the definition of ecosystem services and should be addressed throughout the draft chapter, including here in the framing. This is the only chapter that links changes in biodiversity and ecosystems to the flows of benefits from ecosystems to people. Other chapters talk about livelihood and economic impacts without citing the flows from biodiversity and ecosystem changes to ecosystem services. In this part of Chapter 7, the mechanistic connection is missing. For example, as noted earlier, the chapter is missing discussion and documentation of effects of climate-induced biodiversity and ecosystem changes on agriculture, coastal protection, recreation, fisheries, timber harvest, and carbon sequestration and storage.
- There is virtually no information provided on trends in ecosystem services, with the exception of those given on page 265 and a few other locations. Some of these topics should be brought forward as part of key messages. The discussion of pollination with the necessary documentation is a rare but useful exception.

A straightforward approach to addressing the gaps noted in this review would be to cite or cross-reference a number of useful studies and findings provided in other draft NCA4 chapters. A few examples are provided here.

Examples from draft NCA4 Chapter 9, "Oceans and Marine Resources":

- Page 339, line 24 to page 340, line 11. "This means that fishing communities in Hawai'i and the Pacific Islands, the Caribbean, and the Gulf of Mexico are particularly vulnerable to climate-driven changes in fish populations. Declines of 10%–47% in fish catch potential in these warm regions, as compared to the 1950–1969 level, are expected with a 6.3°F (3.5°C) increase in global atmospheric surface temperature relative to preindustrial levels (reached by 2085 under RCP8.5) (Cheung, Reygondeau, et al. 2016). In contrast, total fish catch potential in the Gulf of Alaska is projected to increase by approximately 10%, while Bering Sea catch potential may increase by 46% (Cheung, Frolicher, et al. 2016). However, species-specific work suggests that catches of Bering Sea pollock, one of the largest fisheries in the U.S., are expected to decline (Ianelli et al. 2016), although price increases may mitigate some of the economic impacts (Seung & Ianelli 2016). Ocean acidification is expected to reduce harvest of U.S. shellfish; while future work will better refine impacts, cumulative consumer losses of $230 million are anticipated under the higher scenario (RCP8.5) (EPA 2017). The implications of the projected changes in fisheries dynamics on revenue (Lam et al. 2016; Seung & Ianelli 2016) and small-scale indigenous fisheries remain uncertain (Weatherdon et al. 2016). Native Americans depend on salmon and other fishery resources for both food and cultural value and there will be significant challenges to some place-based communities (for example, Krueger and Zimmerman 2009; Ch. 24: Northwest)."
- Page 343, lines 5-10. "Coastal communities are especially susceptible to changes in the marine environment (Colburn et al. 2016; Himes-Cornell & Kasperski 2017), and the interaction between people and the ecosystem can amplify the impacts, increasing the potential for surprises. In the Gulf of Maine in 2012, warm temperatures caused lobster catches to peak 3-4 weeks earlier than usual. The supply chain was not prepared for the early influx of lobsters, leading to a severe drop in price (Mills et al. 2013)."

Examples from draft NCA4 Chapter 25, "Southwest":

- Page 1088, lines 21-25. "The irrigation-dependent agriculture of the Southwest provides half of the fruits, vegetables, and nuts of the entire country (USDA, 2015) and most of the national production of wine grapes, strawberries, and lettuce (Starrs and Goin, 2010). Agricultural irrigation accounts for 70% of regional water use (Cooley et al., 2016; Maupin et al., 2014). Consequently, drought and competing water demands pose a major risk for agriculture."
- Page 1088, lines 37-39. "California has the most valuable ocean-based economy in the country, employing over half a million people and generating $20 billion in wages and $42 billion in economic production in 2014 (NOAA, 2017a)."
- Page 1089, lines 18-23. "Over the last five centuries, many Indigenous peoples in the Southwest have been forcibly relocated onto lands with limited water and resources (Bauer, 2016; Denetdale, 2009; Iverson, 2002). This historical legacy exacerbates the impacts of climate change because Indigenous peoples are restricted to increasingly drier areas. Furthermore, many Indigenous peoples depend on natural resources for cultural and subsistence needs, so climate change can negatively affect material and spiritual health (Norton-Smith et al. 2016)."

- Page 1094, lines 11-17. "Wildfire can pose a threat to people, particularly as building expands in fire-prone areas. Wildfires around Los Angeles from 1990 to 2009 caused $3.1 billion in damages (Jin et al., 2015). Respiratory illnesses and life disruptions from one wildfire north of Los Angeles cost an estimated $84 per person per day (Richardson et al. 2012). In addition, wildfires degraded drinking water upstream of Albuquerque with sediment, acidity, and nitrates (Dahm et al., 2015; Sherson et al., 2015) and in Colorado with precursors of cancer-causing trihalomethane (Hohner et al. 2016)."
- Page 1097, lines 18-21. "Currently, 200,000 people in California live in areas 3 feet (0.9 m) or less above sea level, so much of this population is at risk of losing their homes to inundation by 2100 (Hauer et al. 2016). Storm surges and high tides on top of sea level rise would exacerbate flooding (Griggs et al. 2017)."
- Page 1098, lines 17-19. "Shifts in the timing of Dungeness and rock crab fishery into whale migration season in 2016 contributed to increases in whale entanglements in fishing gear (Chavez et al. 2017)."
- Page 1106, lines 16-18. "The California drought led to losses of more than 10,000 jobs and the fallowing of 540,000 acres (220 000 hectares), at a cost of $900 million in gross crop revenue in 2015 (Howitt et al. 2015)."
- Page 1111, line 22-28. "Access to healthcare, social isolation, housing quality, and neighborhood poverty are also key risk factors for heat- related health impacts (Reid et al. 2012). Urban design strategies to address these risk factors include increasing walkability or bicycle safety and adding trees and other vegetation. These strategies can achieve multiple health benefits, including increasing physical activity, thereby helping residents to maintain a healthy weight, reducing the urban heat island effect, and reducing exposure to harmful air pollutants from vehicles."

CHAPTER 8: COASTAL EFFECTS

Summary

This chapter starts strong with a description of the economic value of the U.S. coastline and a sampling of adaptation efforts from around the country. The information would be strengthened by explaining how the examples were chosen and why. For instance, were they the only ones highlighted in the regional chapters? Though the synthesis of material is a nice approach, the examples selected are not largely representative of the scale and scope of current efforts. Also, fact-checking with the source communities is needed, given some small inaccuracies noted in this section. The chapter includes strong statements on flooding that are justified by the science. The key messages are generally positive and are oriented toward informing response actions. In particular, Key Messages 2 and 3 are effective because they highlight the problem as well as a solution path (adaptation). As the chapter progresses, the tone becomes a bit more opinionated and speculative, which should be carefully reviewed by the chapter authors.

Comments Related to the Statement of Task

Comments on Key Messages

> **Key Message 2:** Fisheries, tourism, human health, and public safety depend upon healthy coastal ecosystems. However, coastal ecosystems are being transformed, degraded, or lost due to climate change impacts, particularly sea level rise and higher numbers of extreme weather events. Restoring and conserving coastal ecosystems and adopting natural and nature-based infrastructure solutions can enhance community and ecosystem resilience to climate change and help ensure the continued health and viability of these environments and our coastal communities. Adapting to degradation of habitat integrity and quality may enhance community and ecosystem resilience and decrease both direct and indirect impacts.

Key Message 2 covers so much information that it is difficult to identify the most important points, lessening the impact of the message. It is suggested that it be broken down and the information be further prioritized in order to make the takeaways much more concise. Deleting the first sentence and the "However" start to the second sentence is suggested, with that information then conveyed solely in the supporting text.

> **Key Message 3:** As the pace of coastal flooding and erosion accelerates, climate impacts along our coasts are exacerbating preexisting social inequities as communities face difficult questions on determining who will pay for current impacts and future adaptation strategies and if, how, or when to relocate vulnerable communities. These questions challenge existing legal frameworks; coastal communities will be among the first in the nation to test climate relevant legal frameworks and policies against these impacts. The answers to these questions will establish precedents that will affect both coastal and non-coastal regions.

Key Message 3 introduces the social equity issue, but does not offer strategies to address this vulnerability, which would make the message more effective. Generally, social equity should be introduced report-wide and include examples for how social equity and climate change threats to vulnerable communities can be addressed.

Comments on Graphics

Figure 8.2 is taken from the National Oceanic and Atmospheric Administration (NOAA) and shows selected coastal effects of climate change in each U.S. region. The Miami Beach mention is not quite correct. It states, "Miami Beach, FL has invested $500 million into raising public roads and seawalls, and improving stormwater systems." Miami Beach is *in the midst* of this multi-year $500 million program. Only $100 million has been spent to date in improved stormwater drainage, raised roads, and seawalls. The work is ongoing, but the reference makes it appear to have been completed. The adaptation efforts cited should be verified to the direct source, not just NOAA, to confirm program and status. It is also recommended that the efforts in Puerto Rico be placed in time context (before or after the 2017 hurricane season).

Notably absent in the Figure 8.2 description of the Southeast is the Southeast Florida Regional Climate Change Compact.[1] This is one of the original local intergovernmental models of collaboration in climate change adaptation and mitigation planning, beginning in 2009. Recent developments include the creation of the 2017 updated Regional Climate Action Plan.[2] Figure 8.2 is also hard to read given the extensive amount of small text and it is suggested that this be formatted more appropriately.

It is unclear that the multi-page table (pages 298-302) is associated with Figure 8.2 because no title, description, or linkage to the figure is provided. The table does not seem to be comprehensive, as the Committee identified many missing examples for at least the Southwest region/California. The Chapter 8 authors should consider revising the table to either be more comprehensive or not including it at all. If the table is retained, it should include the examples mentioned in the regional chapters at a minimum, and be explicit in stating the rationale for the inclusion of presented examples.

Figure 8.3 should reference the City of Miami Beach, not the City of Miami. These are two separate municipalities and the incorrect one is noted.

The discussion on equality versus equity presented in Figure 8.4 is important; however, it seems out of place in this chapter, particularly when compared to the other charts, graphs and maps. The discussion would be better placed in the front sections of the draft NCA4, and interwoven throughout chapters as appropriate.

Comments on Literature Cited

It is recommended that reference to news publications (e.g., *The New York Times* and *Scientific American*) be avoided to ensure accuracy of the information. A better source of information is the city itself in order to avoid misinterpretation or speculation.

Authors should be specific when discussing the NCA3, the NCA4, and other related assessment products (e.g., on page 314, line 11, clarify what is meant by NCA).

The chapter should include discussion of compound flooding (see Moftakhari et al., 2017).

Comments on Traceable Accounts

The traceable accounts are supported by appropriate confidence levels. However, the key messages could be more effective if conveyed in a more positive tone to balance risk with adaptation and mitigation efforts.

Other Recommended Changes

The economic and social value of coastline properties is presented, but a further dive into real estate, insurance, and banking would be beneficial for the audience of this assessment report,

---

[1] See http://www.southeastfloridaclimatecompact.org.
[2] See http://www.southeastfloridaclimatecompact.org/regional-climate-action-plan.

especially for decision-makers. This could also include port infrastructure and associated economic impacts.

For the topic of coastal inundation, inclusion of more near-term projections could provide valuable information to decision-makers and other stakeholders.

Cross-referencing other draft NCA4 chapters, including Chapter 9, "Oceans and Marine Resources," and Chapter 11, "Built Environment, Urban Systems, and Cities," as appropriate would also improve the chapter.

CHAPTER 9: OCEAN AND MARINE RESOURCES

Summary

Chapter 9 is centered around three useful themes that underpin the key messages: projected impacts, opportunities for reducing risks, and emerging issues and research gaps. For the most part, these key messages are well selected and the discussion and findings are reported in a credible, transparent manner. The chapter is well organized and effectively orients the reader around the key conclusions, although there is a somewhat abrupt shift between the tone of writing in the introduction, which seems more appropriate for a general audience, and the more technical tone used in the rest of the chapter.

The chapter could benefit from a more comprehensive treatment of the climate impacts on livelihoods by additional attention to other marine-resources based livelihoods (e.g., aquaculture, transportation, recreation, energy) and better linkage to other relevant draft NCA4 chapters.

Comments Related to the Statement of Task

Comments on Key Messages

The three key messages are important areas of focus that concentrate the discussion on iconic species and ecosystems (coral reefs and sea ice ecosystems), impacts on fisheries and fishing communities, and the interaction of long-term climate change and short-term variation creating conditions for disruptive extreme events in the marine environment. The organization of the information and the literature cited provide appropriate support for the messages and reflect current understanding about observed and projected impacts to the United States.

> **Key Message 2:** The Nation's valuable marine fisheries and fishing communities are at high risk from climate-driven changes in the distribution, timing, and productivity of fishery-related species. Ocean warming, acidification, and deoxygenation are projected to increase changes in fishery-related species, reduce catches in some areas, and challenge effective management of marine fisheries and protected species. Fisheries management that incorporates climate knowledge can help reduce impacts, promote resilience, and increase the value of marine resources in the face of changing ocean conditions.

The points being made in Key Message 2 could be strengthened by integrating a more comprehensive treatment of climate impacts on livelihoods associated with fisheries, including aquaculture and recreational endeavors, in the supporting narrative.

> **Key Message 3:** Marine ecosystems and the coastal communities that depend on them are at risk of significant impacts from extreme events with combinations of very high temperatures, very low oxygen levels, or very acidified conditions. These unusual events will become more common and more severe in the future, and they expose vulnerabilities that can motivate change including technological innovations to detect, forecast, and mitigate adverse conditions.

The content of Key Message 3 and supporting text could include stronger discussion of the linkages between extreme events and coastal systems. Cross-reference to the draft NCA4 Chapter 8, "Coastal Effects," is also suggested.

Comments on Graphics

Chapter figures effectively demonstrate important aspects of the key messages in a clear and consistent manner.

Comments on Literature Cited

The citation of current literature and links to the CSSR provide robust documentation for the discussion and key messages. The examples used are linked to points being made and are effective in the discussion. In general, the messages, discussion, and findings are reported in a credible, transparent manner, despite the somewhat abrupt shift between the tone of writing in the introduction, which is more appropriate for a general audience, and the more technical tone in the rest of the chapter.

Comments on Traceable Accounts

The reported process for gathering data and deciding on key messages and supporting documentation for the chapter is inclusive and contains opportunities for meeting the requirements of the traceable accounts. The information and literature citations in the traceable accounts are linked to the key messages, but the Chapter 9 authors should be consistent in how the reported confidence and likelihoods are ordered within the key messages.

Other Recommended Changes

Risk and adaptation measures are generally addressed in the discussion of the key messages, but examples of approaches that could lead to resilience are not apparent. Chapter 9 should include examples, such as:

- Restoration or protection of natural marine ecosystems and ecosystem impacts on coastal flood protection and improvement in fishery habitat.

- Filtration of land-based runoff that can exacerbate temperature and ocean acidification stresses on coastal species is a widely cited strategy for reducing risk of marine systems to climate change.
- Enhancing life history diversity of species through aquaculture, fishery management, and habitat improvements, which can increase resilience to climate change.
- Investments in local ocean-based communities for more diversified livelihoods (cross-reference with other relevant chapters in the draft NCA4).

Chapter 9 could provide more comprehensive treatment of ocean and marine resources, which would make the chapter content more applicable to a wider audience. This could be accomplished in part by addressing the following concerns:

- Reduce the bias towards treatment of fisheries-related issues and add discussion of climate change impacts on other major coastal resource-based economies, including aquaculture practices, tourism, and energy development.
- Include discussion of mitigation in marine systems, which are substantial due to the carbon sequestration and storage capacity of these systems, by at least providing the reader an entry into that literature or referring to the draft NCA4 Chapter 29, "Mitigation: Avoiding and Reducing Long-Term Risks."
- Bring the integrated nature of ocean systems into the discussion through linkages among habitats, species/foodwebs, and humans interacting with marine systems through physically inter-connected ocean currents and tides. Climate change impacts on one of those inter-connected system components can have important cascading effects on other components that are important to note. This approach would also add more focus on people in the chapter.
- Provide change estimates for other marine/coastal systems, besides coral reefs and sea-ice systems, if possible.

CHAPTER 10: AGRICULTURE AND RURAL COMMUNITIES

Summary

Overall, Chapter 10 appropriately highlights the high climate sensitivity of the U.S. agricultural sector and rural communities. It also illustrates some of the complex interactions between physical climate changes, terrestrial processes, and management practices that affect the impacts of climate on agriculture. The traceable accounts section provides ample evidence for the validity of the chapter's main conclusions on how climate changes over the coming decades will introduce new risks to the sector and that it will likely lead to reduced agricultural productivity in the absence of robust adaptation measures.

The chapter currently suffers from readability issues that make the state-of-knowledge difficult to ascertain in several instances. It could be made more effective through substantial improvements to message clarity. The most beneficial improvements would include: reorganizing key messages around impacts rather than forcings in order to avoid oversimplification of impact pathways and unnecessary redundancy in the text; strengthening discussion of recent trends and knowledge advancements since the NCA3; organizing the progression of paragraphs and sections into a tighter logical flow; and explicating baselines in the agricultural sector against which impacts may be measured (while recognizing that some of these will be moving baselines).

The chapter would also benefit from added or augmented background and science content in some places, as noted in this section of the review report. Related to this, figures and case studies should be more tightly coupled to key messages.

Comments Related to the Statement of Task

Comments on Key Messages

The key messages generally include references to the most important pathways associated with climate change impacts on the agricultural sector and rural communities. However, they currently lead to some oversimplification and redundancy, where significant improvements can be made.

> **Key Message 1:** Reduced crop yields, intensifying wildfire on rangelands, depletion of surface water supplies, and acceleration of aquifer depletion are anticipated with increased frequency and duration of drought.
>
> **Key Message 2:** Challenges to human, crop, and livestock health are increasing due to increased frequency and intensity of temperature extremes.
>
> **Key Message 3:** Rural roads, bridges, and community water supply and sanitation are increasingly being damaged by large rainfall events.

Key messages are currently organized around climate change attributes or forcings: drought (Key Message 1), temperature extremes (Key Message 2), and rainfall intensity (Key Message 3). In reality, these attributes are anticipated to combine to jointly affect agricultural productivity, sometimes in complex ways. For example, yields will be concurrently affected by changes in temperature extremes (heat stress), rainfall intensities (leaching, waterlogging, denitrification), and droughts (water stress). This set of complicated interactions is one of the central challenges of accurately forecasting overall climate change impacts on agriculture, and this point should be reflected in the chapter. As currently written, for example, Key Message 1 intimates that increased drought frequency and duration will be the sole or primary driver of yield impacts. This does not appear to be the chapter authors' intent, given that yield impacts are referenced in multiple sections, but the text could be interpreted this way.

To address these concerns, the Chapter 10 authors should consider reorganizing the key messages around categories of impacts. For example, yield impacts as the subject of a key message statement discussing the most likely net effects of changes to temperature, rainfall, and drought. The advantages would be that messages could be framed in a way that is clear about the projected impacts, but does not oversimplify the pathways and interactions, which would clear up some redundancy in the text (e.g., the weather impact on yield is currently mentioned multiple times in the chapter).

The key messages are reasonably well-reflected in the report findings. The Agriculture Report Finding also includes the following sentence, which is broadly in line with the recommendation for rewrites to the key messages: "While some regions may see conditions conducive to expanded (or alternative crop) productivity, overall, yields from major U.S. crops are expected to decline as a consequence of increases in temperatures, and possibly changes in water availability, diseases, and pest infestations" (page 21, lines 22-25).

Comments on Graphics

Figures are mostly focused on industry background information and some climate-driven trends. It is suggested that the Chapter 10 authors consider rebalancing this set of materials to reflect projected impacts emphasized in the key messages.

There may be some missed opportunities for reinforcing key messages with tables, figures and case study boxes. For example, the current sole case study (and Figure 10.3) is about groundwater level trends of the Ogallala Aquifer, which is tangential and potentially misleading; most readers might reasonably assume that the sole case study in the chapter would highlight a well-documented climate change impact, but the southern Ogallala water declines are mostly associated with the juxtaposition of intensive agriculture and low groundwater recharge.

Comments on Literature Cited

The text generally does not provide enough continuity with the NCA3. Key messages between the NCA3 agriculture chapter and the draft NCA4 are rather different. In some instances, this is clearly appropriate, especially where the scope is expanding to impacts on rural communities. In other instances, it is not clear whether the differences reflect new science since the NCA3, intentional differences in highlighted impacts with approximately the same state of knowledge, and/or an artifact of different author group composition and expertise. In accordance with the goal of sustained assessment, effort should be made to comment on the differences between the NCA3 and NCA4 and on the predominant considerations around deviations between them.

The text also does not sufficiently point out the knowledge advances in the scientific literature in the time period following publication of the NCA3 (approximately 2013 to present). On a positive note, the chapter includes a number of relevant journal references published since the NCA3, but there is not much literature synthesis on how data trends, industry trends, model simulation studies, etc. have advanced the state of the science in recent years. Despite the space limitations of the chapter, this is an important addition worth at least one paragraph.

Comments on Traceable Accounts

The traceable account section provides useful information on the chapter development process and contains a good set of references.

Other Recommended Changes

Given the chapter title and scope, it would be valuable to touch on what makes rural communities different from other populations with respect to their sensitivity to climate change. The portions of the chapter text on this topic essentially focus on heightened prevalence of poverty, along with poverty of historically vulnerable populations. However, there are other distinguishing features for many rural and farm communities that may be helpful to consider vis a vis climate change impacts, such as the tight coupling of personal incomes and local tax receipts (funding to schools, etc.) to commodity prices and farm profitability; the reliance on migrant labor (at least in some regions); and the close proximity to air and water quality issues that arise from intensive farming operations. A case study box on rural communities (or one rural community) would be a welcome addition.

Beyond the key messages, the text suffers from some disorganization that diminishes its readability. For instance, the text does not do enough to separate these inter-related but unique considerations:

- Direct biophysical impacts on crops from changing weather variables (e.g., heat stress effect on yield)
- Indirect effects (e.g., accelerated weed pressures, nutrient stress from increased losses)
- Downstream or cascading effects (e.g., food prices, nutrient leaching, water consumption)
- Adaptation methods and their ability to mitigate/buffer/offset the worst impacts (e.g., changing crop rotations)
- The agricultural sector's role in greenhouse gas emissions and carbon sequestration and storage

All of these considerations are touched on to some degree, but as written, it is up to the reader to separate them out into logical categories in order to keep from conflating them, and any given paragraph in the draft chapter may comment on several of these categories.

The chapter would benefit from greater explication of how baselines for assessing impacts are defined. Many of the claims about observed changes (e.g., "losses" on page 381, lines 1 and 4; "production declines" on line 23; and "sea level rise" referenced on page 383, lines 12-13) lack reference to baseline date(s), making the actual magnitude of changes difficult to discern. Then, when measuring projected changes from baselines, there is some unevenness in how economic estimates are provided, and in general it is not clear from the current text what the authors think are the most important/risk-prone/expensive impact pathways among the many that are listed in the chapter. Is it possible to compare projected risks or impacts across the different key messages on economic or other quantitative terms? Could current best estimates be tabulated for quick comparison, even at just an order-of-magnitude precision? If so, this would be a helpful reference for the reader. If not, the NCA4 authors should consider adding a sentence stating that current projection skill does not allow for force ranking in that way.

The "State of the Agriculture and Rural Communities Sector" section of Chapter 10 seems like a good place to add context for readers who are not already intimately familiar with the baselines. Consider adding to this section: (1) quantitative comments in the first paragraph on food security importance in addition to economy and jobs; (2) a big-picture overview of the key products and regions of U.S. agriculture; (3) comments on some key industry trends; and (4) a description of the most important takeaways relevant to agriculture from the draft NCA4 Chapter 2, "Our Changing Planet."

The chapter would benefit from a more logical organization from section to section and paragraph to paragraph. It would be helpful to use section breaks and paragraph separation to clearly separate discussions of different considerations, such as physical climate changes, direct impacts, indirect impacts, downstream impacts, and adaptation measures. Similarly, separation of comments on past (observed) versus future (forecasted) patterns is needed when evidence is referenced.

There are many parts of this chapter that would benefit from a re-write. The first sentence of each paragraph should be a topic sentence that states the main point of the paragraph. As an example, the paragraph beginning on page 374, line 21, starts with "Observed climatic changes are consistent with those predicted by global climate models." The next few sentences might be expected to focus on supporting evidence and details about this point, but instead, the next two sentences are about how agriculture is highly climate-sensitive. Then, the final sentence is about adaption happening up to the present date. These three strands do not flow well together and none of them are well-developed with supporting details.

Related to this organizational suggestion, the NCA4 authors should reconsider the order of topics presented. For example, there is a long discussion of adaption methodologies for industry prior to description of the climate changes themselves and the impact potentials that are motivating the adaptation. This order seems backwards and potentially confusing for the reader.

Some additional specific areas of improvement that are addressed in line-by-line comments (see Appendix B of this review report) include:

- Further development of discussion of land use change as an adaptation mechanism, which is acknowledged in the text but not addressed with much detail or any examples
- Impacts of climate change and carbon dioxide accumulation on nutritional quality of crops, which may be an emerging research area to note
- Impacts on pathways of off-farm environmental impacts beyond runoff and erosion
- Some synthesis comments of major modeling and experimental efforts since the NCA3, either in the body text or traceable accounts section

References to consider including in this chapter, along with other detailed comments, are provided in Appendix B.

CHAPTER 11: BUILT ENVIRONMENT, URBAN SYSTEMS, AND CITIES

Summary

The United States builds slightly more than $1 trillion (2017) worth of buildings and infrastructure per year (U.S. Census[3]) that are largely not designed for a changing climate. Chapter 11 is the home chapter for infrastructure in the draft NCA4 and while it appropriately covers many of the impacts of climate change likely to exacerbate existing challenges in the built environment, urban systems, and cities, there are important areas where mention or expanded treatment would strengthen the chapter and inform readers of the challenges that this broad sector is experiencing. Some areas where expanded treatment is needed include discussion of the non-stationarity in hazards for their use in planning and design practices, consideration of

---

[3] See https://www.census.gov/construction/c30/c30index.html.

multiple climate hazards (or multi-hazards), and risk-informed frameworks of adaptive management. The impact of urban heat islands should also be more comprehensive and perhaps included in a key message. Chapter 11 could also benefit from cross-referencing to the draft NCA4 Chapter 17, "Sectoral Interdependencies, Multiple Stressors, and Complex Systems," among others.

Comments Related to the Statement of Task

Comments on Key Messages

The selected key messages are appropriate, generally clear, consistent, and communicated appropriately for the intended audience.

> **Key Message 2:** Damages from extreme weather events demonstrate current urban infrastructure vulnerabilities. With their long service life, urban infrastructure must be able to endure a future climate that is different from the past. Forward-looking design provides a foundation for reliable infrastructure that can withstand ongoing and future climate risks.

Key Message 2 is a logical outcome from materials covered in the chapter; however, it should mention the lack of current building standards to account for non-stationary hazards in planning and design practices.

> **Key Message 3:** Interdependent networks of infrastructure, ecosystems, and social systems provide essential urban goods and services. Damage to such networks from current weather extremes and future climate will affect many areas of urban life. Coordinated efforts across local, state, and federal jurisdictions to address interconnected vulnerabilities can build urban resilience to climate change.

Key Message 3 should include interdependencies among the hazards that could exacerbate the impacts. Multi-hazard treatments are necessary to examine such effects, such as the dependence of stagnant air with heat waves, that might lead to persisting poor air quality (e.g., Lombardo and Ayyub, 2015).

> **Key Message 4:** Cities across the United States are leading efforts to respond to climate change. Urban adaptation and mitigation actions can affect current and projected impacts of climate change and provide near-term benefits. Challenges to implementing these plans remain, but cities can address these challenges by building on local knowledge and joining multi-city networks.

Key Message 4 should directly call out the use of risk methods to inform policy and decision-making practices for achieving energy and economic efficiencies in solutions or actions. The key message calls out the benefits of multi-city networks, yet fails to showcase any successful models, such as the Southeast Florida Regional Climate Change Compact,[4] which has matured since its mention in the NCA3. Also noteworthy is the 100 Resilient Cities Network,[5]

---

[4] See http://www.southeastfloridaclimatecompact.org.
[5] See http://www.100resilientcities.org.

which was created specifically to address urbanization, globalization, and climate change. The use of these examples in the text supporting this key message would strengthen its impact.

Comments on Graphics

Overall the graphics are clear, internally consistent, and communicated well for the intended audience, with some exceptions noted here.

Figures 11.1, "Current and Projected U.S. Population," and 11.2, "Projected Changes in the Number of Very Hot Days," are appropriate to include, but Figure 11.2 should be shown prior to Figure 11.1. Figure 11.1 would be improved by having the years listed on the figure instead of in the caption. The "Number of People" legends should also be revised to provide consistent binned values and the same colors associated with each binned value in all panels.

The message that Figure 11.3 intends to communicate is unclear. The figure might be improved if the caption were enhanced to provide details that correspond to the three illustrations shown in the figure.

Figure 11.4 is generally effective, but the schematics are unclear. Providing details in the caption that correspond to the items shown in the graphic could help address this.

The chapter would benefit from an additional figure that is similar in format to Figure 11.2, but illustrates projected change in precipitation in urban areas.

The photograph on page 425 should be numbered and the caption enhanced to provide greater rationale for its inclusion in the chapter.

Comments on Literature Cited

The chapter reflects the peer-reviewed scientific literature, with a particular focus on recent literature since publication of the NCA3. No critical content areas are left out of the chapter, although the following items should be included.

Applicable engineering standards require updating so that they provide guidance on computing design extremes of hazards based on non-stationary stochastic processes. Additionally, there is also a need to use new planning and design philosophies, such as adaptive design, observational methods, adaptive risk management, etc. (Wright et al., 2013; Ayyub and Wright, 2016). These efforts are being expanded further by the American Society of Civil Engineers Committee in the ongoing development of a manual of practice on adaptive design and risk management on the adaptation to a changing climate, with a report titled, "Climate Resilient Infrastructure."

Increases in salinity, temperature, and humidity due to a changing climate could result in an increase in corrosion and degradation rates, reducing life expectancy of the built environment and jeopardizing integrity, efficiency, and safety. These effects are not fully explored in the chapter.

Comments on Traceable Accounts

The findings for Chapter 11 are documented in a consistent, transparent, and credible way. They reflect supporting evidence and assessment of confidence levels, although the chapter does not include an assessment of likelihoods, which should be added where appropriate.

Other Recommended Changes

The chapter could be improved by increased treatment of some topics and inclusion of a few additional ones. The Committee recognizes that there are space constraints and that extensive expansion is infeasible, but recommends prioritizing the inclusion of topics that the Chapter 11 authors deem to be effective additions. In revising Chapters 11 and 12, there may be opportunities for the author teams to expand on issues in one chapter or the other and cross-reference. Some issues to consider:

- Expanding the coverage of social and socioeconomic systems and common causal factors across sectors and regions, e.g., consumption behavior, culture, etc.
- Expanded attention to urban transportation systems and underground structures, as they are vulnerable to local flooding, sea level rise, and storm surges.
- Inclusion of impacts of localized flooding due to extreme precipitation.
- More in-depth discussion of urban heat islands is needed, including their magnitude, with reference to the substantial body of literature on this subject. Given the strong effects of heat islands in the built environment, the Chapter 11 authors should consider whether inclusion of this topic in a key message is warranted. Reference to the description of urban heat islands in the draft NCA4 Chapter 5, "Land Cover and Land Use Change," is also recommended.
- Inclusion of monitoring and control needs resulting from the complexity associated with a system of systems, and non-stationarity of interdependencies.
- Expanded discussion of risk, adaptive design, and "Real Options" (e.g., Woodward et al., 2013) for the purpose of allocating resources effectively to achieve economic efficiency.
- More examples and expanded discussion of the integration of natural ecosystems into city planning and design, such as the value of dunes, wetlands, and mangroves.

Finally, improved linkage to the content of other chapters should be considered. This includes cross-reference to the draft NCA4 Chapter 17, "Sectoral Interdependencies, Multiple Stressors, and Complex Systems," as well as Chapter 8, "Coastal Effects," and others as is appropriate.

CHAPTER 12: TRANSPORTATION

Summary

Chapter 12 is well written and appropriately covers the transportation sector as a backbone of economic activity for mobility and connecting key elements of the economy. The chapter is effective in conveying the message that the ability of the transportation sector to perform reliably, safely, and efficiently is undermined by a changing climate due to hazards such as heavy precipitation, coastal flooding, heat, and changes in average precipitation and temperature. These factors impact individual assets across all modes of transportation, affecting the performance of associated transportation networks and imparting critical ramifications to the economy and society at large.

The chapter addresses key elements associated with climate change impacts on the transportation sector, but could be enhanced by including or expanding the discussion of some key issues. These include mention of potential impacts of disruptive or transformative technologies, such as automated vehicles or autonomous aerial vehicles, and the inability of building standards to account for non-stationarity in hazards in planning and design practices. The chapter would benefit from increased discussion of social and socioeconomic systems as they relate to climate change impacts on transportation, and relationships to human behavior, culture, and other factors.

Comments Related to the Statement of Task

Increased coverage of social and socioeconomic systems, and common causal factors across sectors and regions, such as consumption behavior of people, culture, etc. is needed. It is essential to examine the connections among transportation, social, and other systems, and their interdependencies in order to enhance the management of societal responses to a changing climate. These interdependencies are discussed in the draft NCA4 Chapter 17 and cross-referencing might be appropriate, as these systems are also subject to other stressors, such as population growth, economic demands, and technological changes and their potential compounding effects.

Comments on Key Messages

> **Key Message 1:** A reliable, safe, and efficient U.S. transportation system is at risk from increases in heavy precipitation, coastal flooding, heat, and other extreme events as well as changes to average precipitation and temperature. Over the coming decades and the rest of the century, climate change will continue to pose a risk to U.S. transportation performance with differences among regions.

Key Message 1 should be appropriately qualified to account for potential impacts of disruptive or transformative technologies or behavioral changes of users, such as the use of autonomous vehicles or alternative fuel vehicles. These factors could change the use allocation of energy sources with impacts on the environment and related attributes that may then affect hazards and extremes. In the first sentence of Key Message 1, "system" is plural in the "Executive Summary" section and singular in other sections of the chapter. Text should be updated so that identical language is used in all locations where this key message appears.

> **Key Message 2:** The performance and service of the Nation's transportation network is critical for the economic vitality and population mobility across urban and rural landscapes. Extreme events that increasingly impact the transportation network are inducing societal and economic consequences, some of which disproportionately affect

> vulnerable populations. In the absence of intervention, projected changes in climate may lead to increasing transportation challenges, particularly for urban areas because of system complexity, aging infrastructure, and dependency across sectors.

Key Message 2 is a logical outcome from materials discussed in the chapter. However, it should mention the deficiency in building standards to account for non-stationarity in hazards in planning and design practices.

> **Key Message 3:** Engineers, planners, and researchers in the transportation field are showing increasing interest and sophistication in understanding the risks that climate hazards pose to transportation assets and services. Practitioner efforts demonstrate the connection between advanced assessments and implementation of adaptive measures, though many communities still face challenges and barriers to action.

Key Message 3 should call out for the use of risk methods to inform policy and decision-making practices for achieving economic efficiencies in solutions or actions. Although additional research is needed to support broad conclusions and system-wide risk assessments, the value of this type of information may be appropriate to indicate.

Comments on Graphics

Figure 12.1 is likely to be confusing to a general audience. The year 2010 has already passed, so the three maps corresponding to 2010 should be based on the data, i.e., all three maps for 2010 should look the same. The inclusion of 2010 as a projection and why it differs among intermediate-low, intermediate, and extreme scenarios should be explained.

The chapter would benefit from figure(s) showing the broader impacts of multi-hazards on the transportation sector. A good example is Figure 5.1 or 5.3 in the transportation chapter of the NCA3 report (Schwartz et al., 2014).

Comments on Literature Cited

The coverage of specific aspects of the transportation sector is lacking and is noted in this section of the review. Many of these suggestions apply more broadly to the built environment and are also emphasized in the review of Chapter 11. In revising Chapters 11 and 12, there may be opportunities for the chapter author teams to expand on these issues in one chapter or the other and cross-reference.

The Chapter 12 authors should consider including discussion of the challenges posed by the use of engineering standards that do not account for climate change and therefore do not provide guidance on computing design extremes of hazards based on non-stationary stochastic processes. The transportation sector constitutes a significant portion of U.S. investment in infrastructure and, like noted in the review of the draft NCA4 Chapter 11, this infrastructure is largely not designed to standards reflective of the changing climate. Applicable engineering standards require updating so that they provide guidance on computing design extremes of hazards based on non-stationary stochastic processes.

Other areas of recommended expansion of discussion include vulnerabilities of metrorail transit systems to local flooding, sea level rise, and storm surges. Additionally, increases in salinity, temperature, and humidity results in an increase in corrosion and degradation rates that reduce life expectancy of infrastructure and lead to other integrity concerns. As can be adequately supported by available literature, coverage of impacts of future transportation-related technologies, such as alternate fuels as a result of electric cars and autonomous systems, as well as changes in transportation modes and behavior of users and traffic would strengthen this chapter. For Chapter 12 in particular, the importance of taking into consideration changes in human behavior and lifestyle are important, but largely absent in the draft text.

Finally, this chapter would benefit from providing conceptual guidance on performing risk analysis and risk management in order to account for how impacts interact across sectors and scales. Risk analysis and management informs decisions for allocating resources effectively to achieve economic efficiency and could be explored (e.g., Ayyub, 2014).

Comments on Traceable Accounts

The chapter findings are documented in consistent, transparent, and credible terms. They reflect supporting evidence and assessment of confidence levels, although they do not include an assessment of likelihoods, which should be included, when appropriate.

While the chapter cites scientific literature on climate change, it is sparse on scientific literature on climate change impacts affecting transportation. This is acknowledged in the traceable accounts section for Key Message 3, but most references in traceable accounts for all key messages in this chapter are to news reports or gray literature. This may reflect the state of the science on transportation impacts; if peer-reviewed literature on the topic is limited, it could be noted explicitly. The citations are quite useful for establishing previous impacts, although they do not establish deviations from previous conditions or provide insight on attributions, such as separating climate effects from other interdependencies relating to non-climate stressors like aging infrastructure, population, and land-use change. The NCA4 authors should consider whether such an attribution analysis would lead to assigned confidences being overstated.

CHAPTER 13: AIR QUALITY

Summary

Chapter 13 is a new chapter in this NCA and provides useful information on the impacts of air quality on climate change. It is well written and appropriately evaluates scientific evidence on this topic. The technical level is appropriate for a broad audience and effectively conveys the key messages. Recommended improvements include framing climate-related air quality concerns in the broader context of air quality and focusing more strongly on air quality consequences for human health.

Comments Related to the Statement of Task

Comments on Key Messages

For the most part, the key messages are clear, consistent, appropriate, and reflect current understanding regarding the observed and projected impacts of climate change on air quality. However, some adjustments and clarifications may be useful.

> **Key Message 1:** Climate change is increasing the risk of adverse respiratory and cardiovascular effects, including premature death, due to higher concentrations of air pollutants in many parts of the United States. Increased air pollution will also have other environmental consequences, including degraded visibility and damage to agricultural crops and forests. Climate change is promoting weather conditions that more frequently lead to the buildup of ozone and particulate matter and enhance emissions that form these pollutants. These adverse impacts of climate change will compromise ongoing efforts to improve air quality by controlling air pollutant emissions from human activities. Mitigating climate change will also lessen its negative impact on air quality and health.

Key Message 1 notes that adverse impacts of climate change will compromise ongoing efforts to improve air quality by controlling air pollutant emissions from human activities alone. It may be useful to note that many millions of U.S. persons already live in areas exceeding the health-based standards for air quality. This is mentioned later in the chapter, but, if included in Key Message 1, would provide important context and links to non-climate stressor interaction with climate impacts.

> **Key Message 2:** More frequent and severe wildfires due to climate change pose an increasing risk to human health through impacts on air quality. Smoke from wildfires will impair visibility in wilderness areas as well as populated regions. More prevalent wildfires are likely to increase the rate at which outdoor recreational activities are canceled because of the health hazard of wildfire smoke.

For Key Message 2, the links between wildfire smoke and health are understated because the key message relates to outdoor activities. The chapter does note that "wildfire smoke increases the risk of respiratory and cardiovascular disease and poses a substantial health burden" (page 296, lines 20-22) and cites the appropriate references. Given this scientific evidence, the key message for wildfires' impact on air quality should be about the consequences for human health, not for recreational activities.

> **Key Message 4:** Many emission sources of greenhouse gases also emit particles and ozone precursors that affect human health. In addition, methane is both a greenhouse gas and contributes to ozone formation. The human health risks from air pollution can be reduced by addressing these common emission sources.

Key Message 4 is appropriate, but it could be strengthened and clarified by specifically stating that the human health risks in the short term could be reduced by improved air quality in the short term from reduced greenhouse gas emissions. This issue, often referred to as "co-impacts" (or "co-benefits"), has been discussed in the scientific literature and in the draft NCA4 Chapter 14. Some examples are West et al. (2013) and Gao et al. (2018).

Comments on Graphics

The graphics are clear, consistent, appropriate, and reflect the supporting evidence.

Comments on Literature Cited

Overall, the report accurately reflects the peer-reviewed scientific literature. The text on temperature-ozone associations—noting that the effect of short-term exposure of ozone varies by temperature—may be a bit overstated, as it focuses only on one national and one regional study.

The scientific literature on air conditioning is about the prevalence of air conditioning, not the use of air conditioning. This does not change the overall meaning of the text, but should be corrected to be accurate.

Comments on Traceable Accounts

Overall, the findings in Chapter 13 are documented in a consistent, transparent, and credible manner.

In the "Major Uncertainties" section of the traceable accounts for Key Message 1, the text states, "The model simulations that project net increases in future ozone levels over the United States with warmer global climate scenarios have variability in the magnitude of the signal as well as the potential regional differences of the climate impacts on ozone across the United States" (page 500, lines 9-11). This section could note that it is the magnitude and spatial patterns that have uncertainty, not the direction of the signal. It would also be worthwhile to be specific that this refers to tropospheric ozone.

Other Recommended Changes

The chapter notes many adverse health outcomes from poorer air quality, but neglects some other health outcomes that are noted to have associations with air quality, perhaps with growing evidence compared to other health outcomes for which there exists strong evidence. An example is birth outcomes (see Green et al., 2005; Warren et al., 2017). These could be added, as well as text to note other health outcomes could be affected, to give a better indication of the overall health effects of changes in air quality from climate change.

CHAPTER 14: HUMAN HEALTH

Summary

Chapter 14 appropriately evaluates scientific evidence on the human health consequences of climate change and is written at a technical level that is appropriate for the intended audience. The key messages are generally effective and highlight major impacts of climate change on human health. Primary recommendations to improve the chapter include placing greater emphasis on the extent of the impact that climate change is expected to have on human health and expanding discussion of the types of health outcomes that are expected to be impacted by climate change. The inclusion of economic impacts in the key messages should also be reconsidered in the context of the treatment of this topic in other chapters. Most chapters do not report economic information while this chapter does. Generally, the Committee suggests that the NCA4 authors better integrate economic estimates throughout the draft report (see the "Front Matter: Report Findings" section in Chapter 3 of this review report). The NCA4 authors should revise the economics discussion in Chapter 14 to be consistent with treatment of this issue in the report as a whole.

Review Comments Related to the Statement of Task

Comments on Key Messages

Key messages are generally clear and consistent, but some small changes would strengthen their impact. The chapter authors should also review the Health Report Finding provided in the draft NCA4, which the Committee found to be particularly effective in conveying a strong, succinct message about health impacts.

> **Key Message 1:** Although every American is vulnerable to the health impacts associated with climate change, risks are not experienced equally, with older adults, children, low-income communities, and communities of color among the population groups that are particularly vulnerable. Health risks arise from exposure to heatwaves, floods, droughts, and other extreme events; from vector-, food-, and water-borne infectious diseases; from changes in the quality and safety of food and water; and from stresses to mental health and well-being. The risks are projected to increase with additional climate change.

Key Message 1 has two points that may be clearer if divided. The second and third sentences convey that climate change is anticipated to have major and substantial impacts on human health. This is the main subject of the chapter and could be the first key message. The first sentence currently provides a separate message—that some populations are more vulnerable than others. This should either be a separate key message, or listed after the first message of health impacts in general. Within the main text, there should be references provided for the health outcomes listed, including mental health.

> **Key Message 3:** By the end of this century, reducing the severity of climate change by reducing greenhouse gas emissions could save thousands of lives each year and produce hundreds of billions of dollars in health-related economic benefits each year, compared with following a pathway of higher greenhouse gas emissions.

While Key Message 3 makes a valuable and relevant point about the economic impact of the health-related consequences of climate change, it seems a bit uneven that economic estimates are provided in the human health chapter and not in other chapters of the draft NCA4. Furthermore, it is unclear how this value was estimated because the method and data for this estimate are not included.

Comments on Graphics

The authors should consider graphics that align better with the key messages. The figure on the locations of hospitals in flood map regions (Figure 14.2) is not the most compelling figure to convey the consequences of human health in the context of climate change. A figure with a clearer connection to health outcomes would likely be easier to interpret for many readers.

Comments on Literature Cited

Chapter 14 accurately reflects the peer-reviewed literature. More information on other health outcomes would be useful. See "Other Recommended Changes" section.

In the text discussing vulnerable communities, the chapter notes that climate change's effects on health will not be felt equally. However, the chapter neglects to mention that these effects are not felt equally in the present day either.

Comments on Traceable Accounts

The text discussing confidence and uncertainty would be more helpful if it specified which aspect of the key message has uncertainty (e.g., is the magnitude uncertain but the direction of change certain?).

Comments on Data and Analyses

The data and analyses are handled in a consistent manner with the possible exception of the economic results for which the underlying analysis and methods are not presented or well cited.

Other Recommended Changes

The chapter focuses on several of the key health outcomes that are important for climate change, but there are other health outcomes that are likely to be affected. The chapter should mention these as well, even if the level of certainty differs by health outcome (e.g., birth outcomes and lost school days from changes in air quality). As appropriate, this information should be cross-referenced with other chapters, such as the draft NCA4 Chapter 14, "Air Quality."

More explicit information on co-impacts (often referred to as co-benefits) could be included, perhaps as a key message of its own or in relation to Key Message 2, which discusses adaptation policies.

CHAPTER 15: TRIBAL AND INDIGENOUS COMMUNITIES

Summary

It is an enormously challenging task to cover climate change on tribal lands due to the inherent diversity and complexity of Indigenous peoples, their relationships to the environments where they reside, and their legal and political positions within the American system of governance. This challenge is exacerbated when "western science" does not acknowledge the contribution of Indigenous holistic worldviews. Despite this challenge, the chapter does an admirable job of addressing many of the key climate change impacts on Indigenous Peoples based on the available literature. It provides a strong Indigenous voice for climate change science by emphasizing issues that affect Indigenous people and by describing the unique knowledge and perspectives that Indigenous Peoples bring to the issue of climate change.

The Committee's main concern is that Key Message 3 might be misinterpreted by some readers to mean that Indigenous Peoples have such high adaptive capacity, resilience, and experience with climate impacts that they will not be strongly affected by climate change; that climate change represents only an obstacle to their ongoing adaptation. As presently stated, Key Message 3 raises the possibility that Indigenous peoples might actually be less vulnerable to climate change than are other segments of society because of their effective experience in dealing with climate variability. The Committee doubts that the authors mean to imply this. Alternatively, if that is their intention, then they should explain this more clearly. The bottom line is that it is important that the meaning and intent of Key Message 3 is clear. The draft NCA4 Report Finding 10 about Indigenous Peoples in the report front matter is a good example of one way that the issues might be rephrased.

Review Comments Related to the Statement of Task

Comments on Key Messages

The key messages reflect current understanding, but some specific changes are suggested.

> **Key Message 1:** Climate change threatens Indigenous peoples' livelihoods and economies, including agriculture, fishing, forestry, recreation, and tourism. These activities rely on water, land, and other natural resources, as well as infrastructure and related human services that are adversely impacted and will be increasingly impacted by changes in climate.

Although Key Message 1 is clearly written, it emphasizes general statements about climate-change impacts rather than providing examples (and associated references) of the major types of vulnerabilities experienced by Indigenous Peoples. In what specific ways are Indigenous Peoples particularly vulnerable?

The Committee would encourage greater specificity in identifying how Indigenous livelihoods and economies are adversely impacted by climate change. Climate impacts on general sectors, resources, and services mentioned in the key message would affect non-Indigenous as well as Indigenous Peoples. Are there ways in which Indigenous Peoples use resources and services that would cause them to be differentially affected? Although the chapter's lack of specificity is most problematic with respect to Key Message 1, the Chapter 15 authors should consider this same issue with respect to the entire chapter in both the main text and traceable accounts.

> **Key Message 2:** Climate change adversely affects cultural identities, food security, and the determinants of physical and mental health for Indigenous peoples and communities through disruption of interconnected social, physical, and ecological systems.

The supporting text for Key Message 2 might mention (perhaps on page 556 in the paragraph beginning with "Imagery and reports…") that indigenous respect for lands and waters causes climate change impacts on the natural world to have direct psychological impact on indigenous peoples.

> **Key Message 3:** Many Indigenous peoples have been proactively identifying and addressing climate impacts; however, many communities face obstacles to adaptation, including limited capacity to implement adaptation strategies, limited access to traditional territory and resources, and limitations of existing policies, programs, collaborations, and funding mechanisms. Successful adaptation in Indigenous contexts leverages Indigenous knowledge, resilient and robust social systems and protocols, and a commitment to principles of self-determination.

Key Message 3 is ambiguous as to whether Indigenous Peoples are so experienced and resilient in addressing climate impacts that they may (or may not) be particularly vulnerable. Perhaps this ambiguity could be minimized by beginning the message with a statement about obstacles to adaptation, and then pointing out that Indigenous Peoples have a history and unique knowledge that will be valuable in addressing this vulnerability, if the obstacles can be removed. The commitment to self-determination seems important in this context. The Committee suggests that this key message be rephrased to first explain why Indigenous communities are vulnerable to climate change (limited capacity to plan and implement adaptation strategies, limited access to traditional territory and resources, and limitations of existing policies, programs, collaborations, and funding mechanisms—as stated in the key message). Then, the message should state that successful climate adaptation will require full engagement and self-determination of Indigenous Peoples and they must draw upon Indigenous knowledge, culture, and experience in addressing the climate changes that affect them.

Boxes for Key Messages 1 and 3 or cross-references to case studies described in other chapters are two possible ways of increasing these details.

"Key Message 4" is mentioned repeatedly in the chapter but is not included among the key messages, text, or traceable accounts. This discrepancy should be corrected.

Comments on Graphics

Figures would be more effective if they mapped closely to key messages. Figure 15.1 addresses Key Message 3 and Figure 15.2 addresses infrastructure elements of Key Message 2. It would be helpful if Figure 15.2 could be broadened to address the health and cultural issues emphasized in the key message more generally. A figure that illustrates Key Message 1 would also be a welcome addition. A figure analogous to Figure 24.2 in the draft NCA4 Chapter 24, "Northwest" could possibly be used to show how multiple climate impacts affect Indigenous economies and livelihoods (and perhaps also the social and cultural and health issues in Key Message 2).

Comments on Literature Cited

The chapter accurately reflects the recent peer-reviewed literature. No major content areas are missing from the report, but multiple citations are often used to support a single generic statement, without providing details. It would be good to include enough specific details to show how impacts on Indigenous Peoples differ from those on society in general. In some cases, it may be possible to cross-reference examples described in other draft NCA4 chapters, without greatly lengthening the text. The web links in Figure 15.1 do not provide information on tribal adaptation activities, as is claimed in the text. Adaptation activities are described for only three case studies in the third web link provided.

Statements regarding Indigenous Peoples' adaptation to climate change would be more compelling if specific examples were described in the chapter rather than only in cited references. In the "State of the Sector" section, the text states that "the chapter provides evidence that Indigenous people are taking active steps to adapt to climate change" (page 550, lines 16-18). This is repeated on lines 21-22. On page 556, lines 30-31, the text cites two references that document Indigenous People adapting to and coping with climate change, but does not describe the types of adaptation that is occurring. The traceable accounts state that there is robust documentation of ongoing Indigenous adaptation to climate variability and change. Seven references are cited, but no examples are given. There are, however, many examples given throughout the chapter of Indigenous adaptation planning (e.g., page 559, lines 17-23).

The chapter might also mention the importance of outlets other than peer-reviewed literature (e.g., indigenous websites, where they adhere to NCA4 information quality standards) that document insights and information from Indigenous Peoples and give examples.

Comments on Traceable Accounts

The traceable accounts address all of the issues of the key messages, with appropriate citations, but as noted in the "Comments on Literature Cited" section, they are often without specific examples. The uncertainty analysis emphasizes that it is often difficult to project what will happen to Indigenous Peoples with regard to climate change, because studies have not been done to analyze differences in vulnerabilities between Indigenous and non-indigenous peoples. Perhaps this explains why relatively few specific examples of vulnerabilities are presented in the text. Currently, the uncertainty statements in the traceable accounts note few studies on the impact of climate change on Indigenous People. This is inconsistent with the main text, which states that there is abundant evidence that Indigenous People are adapting to climate change.

Other Recommended Changes

The chapter treats many of the key issues about climate change and Indigenous Peoples, but several opportunities are missed that would strengthen the chapter.

Chapter 15 should mention the growing recognition of the value of traditional knowledge in informing place-based impacts of climate change and adaptation, mitigation, and preparedness measures.

Increased awareness among Indigenous People of climate-change impacts and the participation of Indigenous People in landscape-scale initiatives could be highlighted. The fragmented nature of many Indigenous lands suggests the value of partnership of Indigenous People and neighboring landholders. This would provide opportunities to bring traditional knowledge to bear on broad problems.

The chapter should note that there is increased awareness since the NCA3 of the importance of providing local communities with ready access to information, access, and tools (e.g., Indian Energy Program on Requesting Technical Assistance) to inform tribal leadership of alternative choices. See also the recent Sustainable Science Special Feature: Applying Cultural Evolution to Sustainability Challenges (Brooks et al., 2017) and Rodriguez et al. (2017). It would be helpful to specify the progress that has been made since the NCA3 in addressing climate-change effects on Indigenous Peoples.

The potential role of tribal lands and resources to contribute to the development of energy independence and sustainable production of clean energy could also be mentioned (Meisen, 2009; Kronk-Warner, 2013).

CHAPTER 16: CLIMATE EFFECTS ON U.S. INTERNATIONAL INTERESTS

Summary

Chapter 16 explores climate change in the context of international interests in a stand-alone chapter, which is a new and welcome addition to the national climate assessment. The key message themes covered in the chapter—impacts on economy, disasters, conflict, shared resources in border regions—are well-chosen and literature cited is appropriate and current. The analysis of impacts and implications is a bit scattered and not very quantitative, at least in part because data may be few or hard to obtain. Cross-border issues with Mexico could receive more robust treatment.

As a new chapter in the national climate assessment development, it is evident that the scope of this type of chapter is likely still evolving. As written, it is very succinct and could be made more comprehensive and integrative. For instance, it should cross-reference relevant content in the draft NCA4 regional chapters and draw on a broader range of international examples. The international examples currently included in the draft chapter, drawn largely from U.S. Agency for International Development (USAID), are effective. In addition, this chapter could link back to the draft NCA4 Chapter 2, "Our Changing Climate," and appropriate regional chapters. Impacts on agriculture, which should be multi-fold, are only briefly described and would benefit from increased treatment. The discussion of military planning for climate change could also be expanded to strengthen it. For instance, the Coast Guard, with a multi-faceted mission space and many international roles, could be included.

Comments Related to the Statement of Task

Comments on Key Messages

Key Message 2: Climate change and natural disasters can slow or reverse development, undermining investments by the United States in developing countries and increasing the need for additional humanitarian assistance, disaster relief, and even military intervention by the United States. As a response, the United States supports efforts in developing countries to better anticipate and address the impacts of climate change and natural disasters.

The second sentence in Key Message 2 has awkward wording that could be revised to make it more focused. A suggested revision is: "The United States plays an international role in supporting developing countries to better anticipate and address the impacts of climate change and natural disasters." It might also be appropriate to mention and highlight the many agencies in the U.S. government that contribute in this arena. For example, in addition to USAID efforts, there are NOAA and other agency projects under international climate change adaptation and disaster resilience.

> **Key Message 3:** Climate extremes and change, in conjunction with other factors, can exacerbate conflict which has implications for U.S. national security. Climate change already affects U.S. military infrastructure and the U.S. military is incorporating climate risks in its planning.

In Key Message 3, the focus on Department of Defense could be expanded to include the Coast Guard (Department of Homeland Security) and their presence internationally across multiple mission spaces (humanitarian, stabilization, fisheries, search and rescue, etc.) that impact national security. Increased ship traffic resulting from the reduction in Arctic sea ice will also increase security and emergency response responsibilities for the Coast Guard and Navy.

Comments on Graphics

The graphics included in Chapter 16 are clear, consistent, and are communicated appropriately for the intended audience.

Comments on Literature Cited

To the extent it is available, the chapter reflects the peer-reviewed literature. There have been climate assessments for other nations and a selection of these (including Canada) are surveyed in the draft NCA4 Appendix 4. It would be a useful to reference this appendix in Chapter 16 and provide cross-reference to that information, as appropriate.

Comments on Traceable Accounts

Findings are documented in a consistent, transparent, and credible way which reflects supporting evidence.

The treatment of Syria in the traceable accounts could be shortened and improved by mentioning climate extremes (drought/water availability) combined with agricultural practices as factors in the instability there (e.g., Gleick, 2014).

Other Recommended Changes

The chapter could place more emphasis on climate change projections of drying to the South of the United States in Mexico (thought to be "likely") and wetting to the North in Canada (also "likely") and associated impacts. The chapter 16 authors should consider whether more direct links to the Intergovernmental Panel on Climate Change (IPCC) Fifth Assessment Report and relevant information provided in the CSSR (USGRCP, 2017) could be made in some segments of this chapter.

National security discussions appear to focus on "conflict," but other issues could be emphasized. For example, displaced populations, famine, water supplies, global transportation networks, and probably others, could be addressed. Climate change assessments by U.S. military and global security organization(s), such as the Center for Climate and Security, could also receive greater emphasis.

The last few years have seen an increase in the volume of displaced populations, mostly due to regional conflicts. The added challenge of changed or anomalous climate on these populations seems to be a "very likely" stress multiplier, which could be explored (Gleick, 2014).

"Success stories" could receive stronger labeling as such. For example, international weather and observation networks and data sharing via the World Meteorological Organization, international space agencies, and IPCC Assessments. Additionally, the Under2 Coalition—created in 2015 with twelve founding signatories and now has more than two hundred—brings together states and regions internationally willing to make a number of key commitments toward greenhouse gas emissions reduction. The U.S. Climate Alliance, established in 2017, has similar concerns and goals.

CHAPTER 17: SECTORAL INTERDEPENDENCIES, MULTIPLE STRESSORS, AND COMPLEX SYSTEMS

Summary

Chapter 17 provides an overview of the connections among natural, built, and social systems and their interdependencies to enhance the management of societal responses to a changing climate. These systems are subject to a variety of non-climate stressors, such as population growth, changes in economic activity, technological changes, and others, that will have compounding effects. The chapter describes how climate change impacts the stressor dynamics of these systems, their underlying processes, and their interdependencies, and highlights this complexity from a regional and sectoral perspective. Understanding these complexities is critical for effectively and efficiently managing climate risks. Throughout the draft NCA4, the significance of these connections, and interdependencies are evident and this chapter explains and makes more prominent these complexities. It is an important and appreciated addition to this draft fourth assessment.

The chapter adequately provides an overview of the interactions and stressors and is clear, consistent, and communicated appropriately for the intended audience. However, the Committee has some suggestions for chapter improvement. Notably, the text barely mentions the connection to an important system—society. The text should enhance the coverage of social and socioeconomic systems, and common causal factors across sectors and regions. Although it is stressed in Key Message 1, the coverage of societal implications in the chapter is lacking. Examples could also better highlight system-related and climate-related risks and vulnerabilities. For instance, following Hurricane Katrina the power outages meant residents could not get money from ATMs, credit cards did not work, and people could not get paid. People needed money to get their houses repaired and buy supplies, but bank records were often lost and loans could not be secured.

The economic implications for major cross-sector climate change impacts should also be discussed. While "cost" is briefly mentioned in the chapter, the issues associated with economic damages are not. The economic implications can be large and multifaceted and should be included in the text (as highlighted in the Hurricane Katrina example on page 112 of the draft NCA4).

Comments Related to the Statement of Task

Comments on Key Messages

The chapter's key messages are generally clear, consistent and communicated appropriately for the intended audiences.

> **Key Message 1:** Climate change and extreme weather directly impact electricity generation, water supply, food production, human health, social systems (behavior of people—economics, motives, incentives, communities, etc.), and other resources. Traditional approaches to assessing climate change and extreme weather impacts that focus on individual sectors will not yield the needed insights into understanding the interactions within and among these sectors, and how they might be impacted by other stressors. It is not possible to understand the full extent of climate-related impacts on the United States without considering these interactions.

Key Message 1 affirms the understanding that it is impossible to evaluate the full extent of climate-related impacts on the United States without considering interactions, connectedness and interdependencies among systems. Although stressed in Key Message 1, the supporting text should be enhanced to cover social and socioeconomic systems, and common factors across sectors and regions that impact individual behavior of people, including their socioeconomic situation and culture (e.g., Hansen et al., 2013). As written, the section only describes the interaction among "sectors," yet the draft NCA4 as whole places strong emphasis on communities, listing the topic as the draft NCA4 Report Finding 1.

Supporting text for Key Message 1 should also include some discussion of the connection of the impact on communities in the larger system. An example could be drawn from Hurricane Katrina where, because of socioeconomic conditions, adaptive capacity in some communities was limited and evacuation in some areas of the city was hindered. Electricity and communication outages exacerbated outreach and recovery (Zoraster, 2009; Walton, 2015[6]). People were dislocated and infrastructure was damaged, resulting in a decreased customer base for the local power provider Entergy New Orleans, which lost half of its natural gas customers. This eventually resulted in a declaration of bankruptcy by the company.

> **Key Message 2:** Climate change risk assessment requires evaluating how impacts interact across sectors and scales and how they can be shaped by multiple stressors. The

---

[6] See Electric Light and Power, http://www.elp.com/articles/2015/08/ten-years-after-how-entergy-new-orleans-survived-hurricane-katrina.html.

complex risks that result often cannot be fully understood based on any one analysis. Effective assessment of these risks must therefore integrate evidence and explore possible futures, attentive to the ways uncertainties affect decisions and goals.

Key Message 2 is an appropriate and logical outcome based on materials covered in the chapter. Discussion in the section, however, should also mention the need to factor in societal risks in risk assessments. For example, climate change can increase energy demand for cooling, potentially increase air pollution, and increase health impacts. It may be appropriate to cross-reference to other draft NCA4 chapters in the supporting text to convey this information. Framing around social and socioeconomic systems and common causal factors across sectors and regions, such as consumption behavior of people, and culture is also suggested.

> **Key Message 3:** The joint management of interdependent systems can enhance the resilience of communities, industries, and ecosystems to climate change and extreme weather. For example, water resources are often managed to achieve multiple objectives such as flood control, navigation, and electricity production. Such integrated approaches can help avoid missed opportunities or unanticipated trade-offs associated with the implementation of management responses to climate change and extreme weather.

Key Message 3 should directly call out the use of risk methods to inform policy and decision-making practices for achieving economic efficiencies in solutions or actions. The example in the key message should also illustrate a how jointly managing a system can help address climate-related risks. The current example in Key Message 3, while clearly related to climate issues, is not directly about addressing climate risk.

Comments on Graphics

Overall the graphics are clear, internally consistent, and communicated appropriately for the intended audiences.

Figure 17.1 shows changes in water storage in the Southwest, 2011-2013. Although the figure is appropriate for the purposes of Chapter 17, having a figure that demonstrates the complexity of interacting sectors over several regions would better fit with the national-scale topical emphasis of this chapter. Figure 4.2 in the draft NCA4 could be a possible model to adapt for Chapter 17. The Chapter 17 authors could also consider adapting Figure 1 in Steininger et al. (2016).

The chapter uses six boxes to illustrate system complexity, which is more boxes than other chapters. This may be imbalanced, depending on guidelines provided to the NCA4 authors, and the content of the boxes should be carefully considered. Boxes are a useful tool to highlight examples. Flagging key messages in Box 17.1 is also effective. However, all boxes should relate to the chapter topic. Box 17.3 is about wolves. While this may be interesting in some contexts, it is unrelated to climate risk and does not further the readers' understanding of climate interactions. Examples that might be explored to highlight system complexities could focus on the impact of autonomous vehicles on greenhouse gas emissions, vehicle miles traveled and social impacts, or even the implications of a "smarter" grid to improve efficiency. Finally, the boxes illustrate system interdependencies and complexities, but none deal with options and methods that can be used to identify and understand interconnected risks. Expansion to better incorporate risk is encouraged.

Comments on Literature Cited

For the most part, the chapter accurately reflects the peer-reviewed scientific literature, with a particular focus on literature since the NCA3 was published, i.e., since approximately 2013. However, as previously stated, the content on the connection of system impacts on society is not included in this chapter. The Impacts of Climate Change on Human Health in the United States: A Scientific Assessment (USGRP, 2016) could be a useful reference for this topic. In addition, covering this subject matter in terms of an underlying process might be insightful. Specifically, the chapter does not provide conceptual guidance on performing risk analysis and management to account for how impacts interact across sectors and scales for the purpose of allocating resources effectively to achieve economic efficiency (e.g., see Ayyub, 2014).

Comments on Traceable Accounts

The findings are documented in the report in a consistent, transparent and credible manner. They reflect supporting evidence, include an assessment of likelihood, and are effectively communicated.

Other Recommended Changes

The following items should also be mentioned or discussed:

- The economic implications of climate change that causes disruptions across interdependent sectors. These implications can be large and multifaceted and should be included in the text more specifically than as mention of "costs."
- Societal impacts, such as implications of extreme weather events on homeowner/flood insurance and losses or effects of income level on ability to avoid events, as experienced during Hurricane Katrina (IOM, 2007) and Hurricane Harvey.
- Highlight the need for additional monitoring and control technologies due to the complexity associated with a system of systems.
- Discussion of the non-stationarity of interdependencies. This might fit well in the section about unknown implications and impacts.
- The impact of climate change on air quality is especially relevant in the box about wildfires, and should be cross-referenced with the discussion in the draft NCA4 Chapter 13, "Air Quality."
- Although consistent with other chapters in the draft NCA4, using the section heading "State of the Sector" (page 613) is not appropriate because the chapter does not deal with a sector of the economy.
- Guidance or references to practical analytical frameworks should be expanded. While the chapter mentions risk management, no specific tools that a practitioner could use were identified. Some that could be mentioned include:
    - U.S. Climate Resilience Toolkit: https://toolkit.climate.gov
    - International efforts and some are risk specific: https://www.climatelinks.org/resources/climate-risk-screening-management-tool
    - Coastal risks: https://coast.noaa.gov/digitalcoast/tools/flood-exposure
    - Flooding: https://msc.fema.gov/portal
    - Water risks more generally: http://www.wri.org/our-work/project/aqueduct/

CHAPTER 18: NORTHEAST

Summary

The Chapter 18 authors have produced an exemplary chapter. The key messages parallel those in other regional chapters by starting with information on significance of climate change impacts, followed by observations, and then projections of what the future might hold. The inclusion of a key message devoted specifically to adaptation (Key Message 5) is a strong component of this chapter. This is appropriate here because of progress made since the publication of the NCA3. In the NCA3 there were adaptation plans, but little action could be reported across the region. Now, there is action in most states; this evolution should be noted explicitly. Improvements to the chapter could be made by placing more emphasis on urban heat islands and extreme events, among other suggestions provided in this chapter review.

Comments Related to the Statement of Task

Comments on Key Messages

In all cases, including diversity across the region (north to south; large urban areas to rural areas, etc.), the key messages reflect current understanding about observed and projected impacts. The parallel structure of key messages adds to their communicative strength. A few small changes, however, could improve them.

> **Key Message 2:** Many of the services provided by the Northeast's coasts and oceans, including fishing, recreation, and storm protection, are threatened by warmer ocean temperatures, sea level increases, and ocean acidification. Higher ocean temperatures are affecting the productivity and distributions of marine species, and sea level rise is increasing flooding risks. Adaptive capacity of coastal businesses and infrastructure is limited, posing risks to people, species, and economies. Declines in habitats and fishery productivity and increases in sea level would substantially alter coastal landscapes and ways of life in the region.

Key Message 2 seems to contradict Key Message 5 in the statement, "Adaptive capacity of coastal businesses and infrastructure is limited, posing risks to people, species, and economies." This inconsistency should be remedied by changing the language or adding appropriate qualifiers.

> **Key Message 4:** The history, culture, entertainment, government, businesses, and diversity present in the Northeast's urban centers and their interconnections make Northeast cities critical for economic opportunity and innovation. Disruptions to infrastructure and negative impacts on historic sites, health and well-being, and urban economies are already occurring and will become more common with a changing climate.

In Key Message 4, the sentence beginning with "Disruptions" is oddly composed/worded and some focus on historic sites seems out of place. Some clarification or revision to the sentence is recommended.

Key Message 5: A wide range of communities in the Northeast are taking action to reduce the risks posed from climate change to human health, economies, and ecosystems by proactively planning and implementing climate adaptation and enhancing resilience in health, transportation, planning, communication, and other sectors. These communities are using decision support tools to assess risks and vulnerabilities, promote innovative responses, and maintain sustainable and diversified ecosystems, thereby demonstrating the value of workable adaptation solutions by early adoption.

Key Message 5 represents an important advancement in responding to the impacts of climate change. This key message is different from the corresponding message in the NCA3 because it reports new actions undertaken in the last few years. This difference between the NCA3 and NCA4 should be strongly emphasized.

Comments on Graphics

Figure 18.7 should either be replaced with a simpler version—perhaps just the second part—or explained more thoroughly. Two versions of the same map with shaded regions that are not explained is not appropriate for the intended audience.

Comments on Literature Cited

The majority of references have been published since 2013, and their assessment reflects progress in the science since the NCA3. The coverage, being organized by differentiating urban from rural, clearly shows how new knowledge confirms and strengthens conclusions from the NCA3. The new structure of the NCA4 that allows expanded treatment of regions has been well used by the Chapter 18 author team.

Comments on Traceable Accounts

The traceable accounts are strong and could be used as a model for other chapters where improvements to the traceable accounts are suggested. They are clearly articulated and indicate confidence clause by clause in some cases, which is appropriate. The treatment of uncertainties, confidence, and likelihood is also very good.

Other Recommended Changes

More emphasis should be placed on urban heat islands and extreme events that impact the people where they live. There is a box on this topic covering Rhode Island (under Key Message 3 on rural economies), but it is a dominant, more widespread issue that deserves to be covered earlier in the chapter instead of being diluted and covered partially across multiple sections of the chapter.

As noted for the draft NCA4 as a whole, Chapter 18 could be improved by taking a little more care in differentiating what is new from the shorter Northeast chapter in the NCA3. While the draft NCA4 Chapter 18 does a better job than many others overall in the inclusion of recent information, emphasizing the advancement since the NCA3 could be stronger.

CHAPTER 19: SOUTHEAST

Summary

Overall, Chapter 19 does a nice job of conveying key messages for the Southeast region and emphasizing appropriate materials. Much of the chapter focuses on natural landscapes, which are an important component of climate change in the Southeast, but the chapter would benefit from an expanded perspective on urban systems and their adaptation responses. In particular, urban heat islands and their interaction with extreme heat events could be discussed at the regional level. This should cross-reference the draft NCA4 Chapter 5, "Land Cover and Land Use Change," where this topic is also discussed, and Chapter 11, "Built Environment, Urban Systems, and Cities," if the authors follow the recommendation in this review report to increase attention to land cover and land-use change in that chapter.

Increased citation of existing local and city action plans would strengthen this chapter and highlight steps being taken to respond to climate change. Also, port infrastructure at medium-size ports, such as Charleston, South Carolina, is expanding to attract Panamax traffic. This is a major investment in infrastructure that makes these cities more vulnerable to sea level rise and discussion of this could fit in the Charleston case study that is already included in the draft chapter.

Comments Related to the Statement of Task

Comments on Key Messages

Overall, the selected materials and treatment of key messages is appropriate. However, the key messages should be revisited to balance the vulnerability being highlighted with the adaptation effort to address the risk.

> **Key Message 1:** Many Southeastern cities are particularly vulnerable to climate change compared to cities in other regions, with expected impacts to infrastructure and human health. Increasing heat, flooding, and vector-borne disease could affect the vibrancy and viability of metropolitan areas. Many of these urban areas are rapidly growing and offer opportunities to adopt effective adaptation efforts to prevent future negative impacts of climate change.

Key Message 1 would be more clear if the second sentence began with, "In the absence of adaptation."

> **Key Message 4:** Rural communities are integral to the Southeast's cultural heritage and to the vibrant agricultural and forest products industries across the region. Increasingly frequent extreme heat episodes and changing seasonal climates will increase exposure-linked health impacts and economic vulnerabilities in the agricultural, timber, and manufacturing sectors. By the end of the century, over one-half billion labor hours could be lost from extreme heat related impacts.

Many of the impacts listed in Key Message 4 are not unique to rural communities and could also logically be included in Key Message 1. The Committee recommends restructuring the key messages to better reflect similar versus distinct impacts in these different settings.

Comments on Graphics

The graphics and maps in Chapter 19 are generally easy to follow for the intended audiences. In particular, Figure 19.10, showing the October 2015 Extreme Rainfall Event, is a very effective case study that is specific and focused. Table 19.1 is also effective.

Figure 19.14, "Projected Changes in Hours Worked," has large economic impacts. It is a national map and should also be emphasized in either the draft NCA4 Overview Chapter or in a national topic chapter.

The y-axis in Figure 19.8, "Highest Daily Water Levels," is confusing and should be better explained in the caption.

Comments on Literature Cited

This chapter is heavy on literature, yet light on local action plans. It is recommended that the Chapter 19 authors draw more on existing plans to highlight new activities, particularly to show expanded efforts since the NCA3.

Comments on Traceable Accounts

The traceable accounts are thorough and provide appropriate confidence and likelihood designations, except for Key Message 4. Further explanation of the research supporting the statement, "By the end of the century, over one-half billion labor hours per year could be lost from extreme heat-related impacts (medium confidence)" in Key Message 4 is needed.

Other Recommended Changes

Port infrastructure at medium-size ports in the Southeast, such as Charleston, is expanding. This is a major investment to port facilities and with this infrastructure expansion comes more vulnerability to sea level rise; discussion of this could fit in the Charleston case study.

The Southeast Florida Regional Climate Change Compact[7] is only mentioned in passing on page 754 and it is recommended that this example be expanded. This is one of the original intergovernmental models of collaboration in climate change adaptation and mitigation planning since 2009 and could be further emphasized in Chapter 19. Recent developments include the creation of the 2017 updated Regional Climate Action Plan,[8] which could be mentioned or discussed.

[7] See http://www.southeastfloridaclimatecompact.org.
[8] See http://www.southeastfloridaclimatecompact.org/regional-climate-action-plan.

The mention of Miami Beach on page 734 is not quite correct. It states, "Miami Beach, FL has invested $500 million into raising public roads and seawalls, and improving stormwater systems." Miami Beach is *in the midst* of this multi-year $500 million program. Only $100 million has been spent to date in improved stormwater drainage, raised roads, and seawalls. The work is ongoing, yet the text makes it appear to be completed. Adaptation efforts cited should be verified to the direct source, in this case the city, to confirm program and status. This specific topic, and the chapter more generally, should also be linked to the similar discussion in the draft NCA4 Chapter 8, "Coastal Effects."

The Chapter 19 authors should consider elaborating on vector-borne disease as related to climate change (page 725), as this is a growing challenge for local leaders in the Southeast. As appropriate, this should also be linked to relevant discussion in the draft NCA4 Chapter 14, "Human Health."

While the U.S. Caribbean now has its own chapter and voice (draft NCA4 Chapter 20), the authors should consider pointing out common risks and efforts between the Caribbean and the Southeast, particularly southeast Florida, which can have more in common with the islands than with states in the northern part of the region. The two chapters should also be cross-referenced, as appropriate.

CHAPTER 20: U.S. CARIBBEAN

Summary

The U.S. Caribbean Chapter of the draft NCA4 represents a welcome addition to the national climate assessment. Previously, this region was incorporated into the Southeast regional chapter, which made it more challenging to emphasize unique climate impacts and responses occurring in the Caribbean Region. The Committee commends the NCA4 authors on inclusion of this new chapter and provides suggestions for further capturing important attributes needed to understand climate change on these U.S. islands.

The chapter is well written and at an appropriate level of detail for the intended audience. Findings are presented and documented in a consistent manner. The key messages are consistent with current understanding about observed and projected impacts from the perspective of small islands in this region and are clear despite their multidimensional nature. The chapter could do a better job of explaining and balancing information on the two sets of islands in the U.S. Caribbean—Puerto Rico and the U.S. Virgin Islands. These set of islands differ considerably in size and population, and in some cases, they have differing vulnerabilities and risks associated with climate change. The availability of data to inform this report and current actions being taken to address climate change also varies, with Puerto Rico having a much richer set of information to draw from, as is reflected in the chapter. However, where possible, the discussion should be better balanced to present a more complete understanding of climate change effects in the region. It is also suggested that where appropriate, greater cross-referencing to other chapters in the draft NCA4 be made, including to Chapter 27, "Hawai'i and U.S. Affiliated Pacific Islands." Despite differences in status (state versus various U.S. territory designations) and some socioeconomic parameters, these island regions are similar in that there are limitations due to their small size and the tight connections between natural resources, culture, and economic activities. Knowledge gained through evaluation of climate change impacts in one region may therefore help to inform understanding of climate impacts and response actions in the other.

Comments Related to the Statement of Task

Comments on Key Messages

The six key messages focus on critical challenges, emerging issues, opportunities, and success stories for addressing risks related to climate change that are being experienced by the islands and people of the U.S. Caribbean. Despite the paucity of data on the U.S. Virgin Islands, the approach taken paints a complex, but reasonably accurate picture of climate impacts through a multidimensional lens and concludes with a section on the value of adaptive capacity and building resilience.

The key messages are in line with the current understanding about observed and projected impacts to the United States from the perspective and position of small islands in the Caribbean Sea and Atlantic Ocean. They are clear and the supporting text is closely linked to the points being made. The narrative and supporting evidence of the key messages includes good indications of the levels of risk.

> **Key Message 6:** Shared knowledge, collaborative research and monitoring, and building institutional adaptive capacity can reduce the need for disaster relief, enhance food security, and improve economic opportunity in the U.S. Caribbean. International cooperation and strengthening partnerships in the Caribbean reduces vulnerability and can reduce risks associated with climate change uncertainty.

Key Message 6 has been used as a means of addressing the underlying conditions that influence the success of U.S. Caribbean communities undertaking adaptation and mitigation initiatives in response to a changing climate. This acknowledgement is an important one, but associated challenges in sustaining this effort could be mentioned. Key Message 6 would benefit from the inclusion of an example or two to improve understandability for readers.

Comments on Graphics

In general, the graphics provide strong support for the messages and the narrative.

For Figure 20.1, the caption should say "seven" inhabited islands, not six. Note there are four inhabited U.S. Virgin Islands, including Water Island. The population size difference (3.4 M versus 106K) should also be noted.

In the caption for Figure 20.5, the described positioning of the graphs is different than the layout.

Figure 20.7 has an important message, but it looks more like life in the Pacific than the Caribbean and therefore may not resonate well with the regional audience.

The two photographs on page 809 of the draft chapter have no figure numbers and they contradict or bring into question the statements in the paragraph above (lines 4-8) that state that the U.S. Caribbean has not been hit by a major hurricane in recent years. This text should be updated to reflect the 2017 hurricane season and the discussion of this topic in other chapters in the draft NCA4.

Comments on Literature Cited

The chapter utilizes peer-reviewed literature and includes a noticeable number of studies published post 2013 and the NCA3. Much of the data and information are derived from U.S. government technical agencies' data and reports developed by regional intergovernmental entities and regional technical research groups. The chapter information includes work utilizing climate downscaling to address the resolution challenges small islands have with global and even regional models. This is an important contribution to the understanding of the climate-related data and information associated with the U.S. Caribbean. The expansion of this effort to an even finer resolution that would provide information of greater direct relevance to the Virgin Islands would allow for the picture presented to be more complete (e.g., see Figure 20.3). More generally, expanded discussion of recent science findings from the Virgin Islands is warranted.

The chapter acknowledges that the CSSR does not include data specific to the U.S. Caribbean and tackles the task from a number of information and data sources, including IPCC reports and linking relevant information contained in the CSSR, which support the contentions of the messages.

Comments on Traceable Accounts

Generally, the key messages are presented and documented in a consistent manner and make understandable links between the literature and the likelihood and confidence provided in Key Messages 2-5, although likelihood and confidence statements are not provided in the key messages, as they are in other chapters. The challenge of presenting information for the two sets of islands with very different information bases and conditions makes transparency difficult. There are a number of places in the chapter where it either appears that only Puerto Rico has the noted impact or situation, or that both sets of islands are experiencing the same, which is sometimes not the case. Part of the challenge is that there is often a lack of readily available information from the Virgin Islands. Puerto Rico is 26 times the land size and more than 33 times the population of the U.S. Virgin Islands. Puerto Rico is also significantly more advanced in their efforts to address climate change adaptation as a society than the Virgin Islands, where the level of response to the challenges is notable, but not necessarily well documented at this time.

Key Messages 1 and 6 are not linked to the evidence in a clear manner. In Key Message 1, the focus is on water resources and the negative changes in freshwater availability that are projected. The evidence base and the uncertainties relate to the message, but the confidence statements are linked to a possible impact. It is suggested that more about the impacts be discussed earlier in the text, or a more direct connection be made in the confidence statement paragraph. In Key Message 6, the evidence base and the major uncertainties sections briefly speak to increased adaptive capacity linked to collaboration, joint projects, and shared knowledge. Examples could strengthen these sections. The confidence statements could offer medium confidence for small island states in the U.S. Caribbean being able to develop useful and needed mitigation and adaptation plans because of cooperation, shared data, and collaborating expert resources.

Other Recommended Changes

The chapter authors should consider using the background section to provide greater context and detail about the region, including information that may be unique relative to other U.S. regions, such as the island population sizes, research and expert resource pools, and the absence of constantly running water sources (in the U.S. Virgin Islands). The sentence, "The Caribbean is expected to warm faster than the global average and to experience greater sea level rise than global estimates from the Intergovernmental Panel on Climate Change (IPCC 2012)" (page 812, lines 2-4) would be an appropriate statement for the introduction or background.

The introduction in Chapter 20 should inform the reader about the imbalance of information from Puerto Rico and the U.S. Virgin Islands. This should include identification of unique attributes and commonalities of the island sets that informs discussion of vulnerabilities, risks, and impacts. Some information that could be noted is that the U.S. Virgin Islands have infrastructure and historical buildings in the inundation zone for sea level rise, including the power plants on both St. Thomas and St. Croix, schools, housing communities, the towns of Charlotte Amalie and Christiansted, Frederiksted, and pipelines for water and sewage.

The narrative of the chapter makes it clear that islands concentrate hazards and exposure in a changing climate. This point could be drawn out more fully in the text to differentiate this region from the others, and emphasize commonalities with Chapter 27, "Hawai'i and Pacific Islands."

It is recommended that the inclusion of emissions scenarios (Special Report on Emissions Scenarios [SRES] versus RCPs) be explained, or greater consistency adopted. The introduction to the draft NCA4 informs readers of the RCP scenarios that will be used in the document, but a number of citations in this chapter use SRES (see Figure 20.3) and no explanation is included.

Special care should be taken to identify linkages between this chapter and other relevant draft NCA4 chapters. This includes cross-reference with Chapter 27, "Hawai'i and U.S. Affiliated Pacific Islands," with Chapter 8, "Coastal Effects," including discussion of coral reefs, and the Chapter 29, "Mitigation: Avoiding and Reducing Long-Term Risks," section on "Challenges, Opportunities, and Success Stories for Reducing Risk."

CHAPTER 21: MIDWEST

Summary

The Committee found Chapter 21 to be strong overall. The key messages address main impacts, vulnerabilities, and adaptations in major sectors of concern within the region. Key message structure and language is clear and internally consistent. The chapter is written at an appropriate technical level for the intended audience, striking a reasonable balance between accuracy and scientific content and is generally accessible. The chapter also draws well on current understanding and recent peer-reviewed literature and to the findings reported in the

CSSR. While some minor issues were raised in the review, no major concerns are expressed or substantial improvements recommended.

Comments Related to the Statement of Task

Comments on Key Messages

The key messages are well developed. Each takes the form of presenting the context for the concern within each sector, the impacts of relevant changes in climatic conditions, how these impacts might produce changes of importance to stakeholders in the region, and steps that have been or can be taken to adapt to these changes. This structure makes for relatively long key messages, but provides reasonably complete picture for readers of the changes and their impacts. When comparing the key messages to the chapter's "Summary Overview" section, it is evident that the third paragraph (page 844, lines 24-34) makes significant mention of the importance of and impacts on the Great Lakes in the region, but this is not mentioned in the key messages. The Chapter 21 authors should consider adding a key message on the Great Lakes, or incorporating Great Lakes impacts into Key Messages 3 and 5 (at least).

> **Key Message 1:** The Midwest is a major producer of a wide range of food and animal feed for national consumption and international trade. Increases in warm-season absolute humidity and precipitation have eroded soils, created favorable conditions for pests and pathogens, and degraded quality of stored grain. Projected increases in moisture, coupled with rising mid-summer temperatures, will be detrimental to crop and livestock production, putting future gains in commodity grain production at risk by mid-century.

Key Message 1 makes no mention of adaptation, whereas the others do. Adaptation in agriculture will be critical and a summary of the categories of adaptations available within this sector could be useful here. This is true for Key Message 1 itself, and in the "Summary Overview" section of the chapter (page 844, lines 8-15).

> **Key Message 3**: The ecosystems of the Midwest support a diverse array of wild species, and provide essential services such as water purification, flood control, crop pollination, and recreational opportunities that support human livelihoods. Species and systems are typically most at risk when climate stressors interact with land-use change, habitat loss, pollution, and invasive species. Restoration of natural systems, increases in the use of green infrastructure, and targeted conservation efforts, especially of wetland systems, can help protect people and nature from climate change impacts.

Key Message 3 and its supporting text could do a better job articulating interactions of climate impacts with land use changes, particularly fragmentation and urbanization effects.

Comments on Graphics

The graphics are generally useful. Figure 21.2 is quite effective.

Figure 21.1 provides a nice non-technical introduction to the idea of vapor pressure deficit, but it is questionable whether the included maps are appropriate to include. The Chapter

21 authors should consider whether the same idea (plant stress) could be conveyed with maps of increased actual evapotranspiration (AET) or potential evapotranspiration (PET) or PET minus AET, which may not require the ecophysiology lesson to accompany it.

Comments on Literature Cited

The discussion of impacts of increased moisture in agricultural systems might benefit from further contextualization and discussion of the potential for future drying. The traceable accounts section cites the CSSR: "future higher temperatures will likely lead to greater frequencies and magnitudes of agricultural droughts throughout the continental United States as the resulting increases in evapotranspiration outpace projected precipitation increases." Additional literature supports this possible impact based on the Coupled Model Intercomparison Project, including Douville and Piazotta (2017) and Berg et al. (2017). The authors should consider drawing greater attention to this recent literature suggesting future drying as context, at least for the existing discussion of observed and expected increases in moisture.

The section on adaptation in agriculture (page 851, lines 1-10) is missing reference to two key sets of literature. First, there is some work to begin mapping dynamics in irrigation (Brown and Perez, 2014). Irrigation is a possible adaptation, and it is possible to begin to see and comment on the geographic variation in irrigation as an adaptation strategy. Anecdotally, irrigation is being increasingly used on droughty soils in Michigan as a back up in dry years. Second, the chapter makes no mention of land-use change as an adaptation strategy. Agricultural adaptation is not limited in theory to changes in management practices. Abandonment in some areas and (re) initiation in others is also possible. There is some evidence in the econometric literature for this already happening. For instance, Burke and Emerick (2016) provide evidence of the limits to adaptive response to extreme heat in U.S. agriculture, and cannot rule out abandonment of cropping as a recent adaptation. Feng et al. (2013) show that outmigration from rural areas is partly related to declines in yields.

Comments on Traceable Accounts

Between the main text and the traceable accounts, the chapter does an excellent job of supporting the claims in a consistent, transparent, credible way. The traceable accounts related to human health are slim. It would be useful to ensure that each claim in the human health section is supported with traceable accounts.

CHAPTER 22: NORTHERN GREAT PLAINS

Summary

The Committee is pleased that the Great Plains region has been broken into two separate chapters for the NCA4 instead of continuing with the NCA3 model of having a single chapter to discuss the region. Chapter 22 focuses on the consequences of climate change in the Northern Great Plains, with particular attention on water resources, agriculture, ecosystems and recreation, energy development, and Indigenous People. These are appropriate topics for this geographically diverse, low-population region. The chapter is well written and clear about the impacts of climate change in recent decades and in describing some future adaptation options. It makes good use of the published literature on some topics but not all, probably reflecting the inevitable imbalances of expertise of the author group across this diverse geography and set of topics. The chapter provides a useful baseline context on the physical and economic geography of the region, although some improvements can be made, as noted in this chapter review.

The chapter links to the draft NCA4 Chapter 2, "Our Changing Climate," and provides a helpful and succinct paragraph on the key aspects of climate change projections specific to the region, highlighting warmer average temperatures, reduced average streamflow and snowpack, variable changes to average precipitation, and increasing precipitation intermittency. In this way, the content of the chapter incorporates findings from the CSSR as it applies to regional concerns. Given that there are strong geographic gradients in some of the climate projections within the Northern Great Plains, a follow-on paragraph describing the sub-regional patterns and an associated figure showing regional projections for basic climate variables (e.g., annual and seasonal temperature and precipitation) would be helpful.

The main concerns for this chapter center around: (1) the lack of discussion of the geographic variability of climate and climate projections across the region to recognize the great range of climate conditions; (2) the rather thin discussion of which agricultural impacts related to climate change (as opposed to other factors) influence decision making; (3) the lack of inclusion of national parks and monuments in the discussion of ecosystems and recreation, including the economic challenges to local communities; and (4) the speculative nature of some of the takeaways concerning climate change impacts to energy production. The phrasing of some of the key messages seems ambiguous and phrases like "parts of the region" (where exactly?) or "unprecedented variability" (over what time span?) require more specificity to be meaningful. The chapter makes no explicit linkages to the NCA3, which would help identify updates in information and coverage. It is also noted that none of the authors are from Montana, North Dakota, South Dakota, or Wyoming (four of the five states included in the region), and the largely extra-regional authorship could raise concerns for some stakeholders in these states.

Comments Related to the Statement of Task

Comments on Key Messages

For the most part, the topics included in the key messages reflect current understanding about observed and projected impacts in the region, and they identify widely recognized issues with respect to climate change. However, the wording of some of the key messages was found to be unclear and open to multiple interpretations. Increased specificity is recommended, as detailed in this section.

> **Key Message 1:** Effective water management is critical to ensuring the region has enough water to meet the demands of its people, its crops and livestock, and its energy industry. Even small changes in precipitation can have large effects downstream, and when coupled with the variability from extreme events, makes managing these resources a challenge. Future changes in precipitation patterns and the potential for more extreme rainfall events will only serve to exacerbate these challenges.

Key Message 1 basically suggests that water management is already tricky in the region and any future climate changes will make it even trickier, even if the projected changes in precipitation are relatively modest. The Committee questions whether this is correct. On the point about small climate changes having large impacts on "downstream" effects, it is unclear whether the Chapter 22 authors mean terrestrial hydrology, or industries that depend on them, or both. The message mentions extreme rainfall events, but not drought. Given that this key message focuses on the fact that climate change will make water management more difficult, the text falls short of suggesting that more flood or drought events will cause economic damage and it is unclear whether that a fair interpretation of the message. It is suggested that this key message be rewritten to avoid ambiguity.

For the text supporting Key Message 1, projected changes in precipitation at the national scale (Figure 2.6 in the draft NCA4) show considerable geographic and seasonal variation in precipitation across the Northern Great Plains. Western Montana and Wyoming are projected to become drier whereas the Dakotas and Nebraska will become wetter. In addition, much of the region will experience less precipitation in summer, leading to heightened drought during the growing season. The Committee recommends that Key Message 1 and other statements about projected climate change be specific about the subregion and season being described.

The text supporting the Key Message 1 discussion that climate change will exacerbate existing water management challenges in the region is defensible and in line with the current scientific understanding, and the emphasis in the text on water storage (both as ground and surface water) as a buffer against extremes is appropriate. However, it should be noted that groundwater recharge rates are highly variable, ranging from rapidly recharged floodplain aquifers supported by irrigation (western Montana), to aquifers that replenish on much longer time scales (Madison Limestone Aquifer, Ogallalah/High Plains Aquifer). For the groundwater-irrigated parts of the region, there has been some work on modeling changes to recharge and withdrawals under future climate change that can help assess the most likely impacts on water for food production. These should be referenced. It would be helpful to add case studies from those watersheds that are likely to become the most problematic in this light. See also the "Comments on Traceable Accounts" section of this chapter review.

> **Key Message 2:** Agricultural production in the Northern Great Plains, with gross revenue of $52.3B per year, has benefited from longer growing seasons and other recent climatic changes. Additional production and conservation benefits are expected in the next two to three decades as land managers employ innovative adaptation strategies, but changes in extreme weather events may offset some benefits. Adaptation to longer-term climate changes will likely require transformative changes in agricultural management, including geographical migration of agricultural practices and enterprises.

Key Message 2 states that climate change will improve the region's agricultural industry on average in the coming decades, but should make clear that rising temperatures and extreme events will offset most of the benefits in the long term. The primary challenges to agriculture will come from drought in the western subregion and higher precipitation and flooding on the eastern side, and this should be noted. Seasonality of precipitation is particularly critical for agriculture. Reference to geographical migration of agricultural practices and enterprises is not well discussed in the text and is unclear in the key message. Overall, it is recommended that Key Message 2 be reworded for greater clarity.

The supporting text for Key Message 2 appropriately points out the complex and countervailing trends that will likely determine the net effect of climate change on agriculture in this region. On the one hand, warming in some of the colder parts of the region may lengthen the crop growing season and accelerate crop phenological development. Those short-term changes can potentially benefit productivity of existing agricultural land, as well as open up new land for cultivation potential (although Key Message 3 argues for a cautionary approach to agriculture expansion). On the other hand, rising temperatures, increasing variability in rainfall, and the likelihood of more extreme events are generally detrimental for agricultural productivity. These points should be strengthened in the text.

There are a few other potential mechanisms relevant to Key Message 2 that are mentioned in the draft NCA4 Chapter 10, "Agriculture and Rural Communities," and could be included here. A general suggestion would be to better leverage some of the science referenced in Chapter 10 in order to make a more detailed assessment of the individual potential impacts and how they are likely to come together. Specifically, the conclusion that climate change will benefit agriculture at least in the near term is plausible but not well supported by the limited evidence provided. The Committee suggests that either a stronger scientific case made for this if it exists, or a rephrasing of the text to indicate that there is not enough known to have confidence in whether the net impact on the region will be positive or negative. The Committee agrees with the comments on adaptive management, but a substantially open question is whether farmers can adapt their practices fast enough as future changes accelerate; see the NCA3 agriculture chapter for a good discussion on this topic (Hatfield et al., 2014). Also, the supporting text for Key Message 2 describes some important trends, especially more land being broken out for row crops, but does not explain how much attribution belongs with climate change.

A clearer discussion about the multiple factors that drive producer decision-making with respect to agriculture would provide useful context. This analysis is discussed to some degree in draft NCA4 Chapter 10 and should be cross-referenced in this chapter. At the farm or ranch level, local climate variability is one component among several in the decision- making process (e.g., government policies, insurance, global and local prices, contracts, expected price received, production inputs, pests). Contrary to statements in the text, not all land-use change is a result of climate change and the response of farmers and ranchers will be based on a number of local and global factors.

In general, the agricultural references in Chapter 22 emphasize eastern region crops, with less consideration of dryland winter/spring wheat, hay production, small grain, pulse crops, livestock, and legume/oil seed rotational crops in Montana and North Dakota (see the "Comments on Literature Cited" section of this chapter review for some suggested references). The Matador Range example is a good one for grazing operations in the western subregion, but how that program relates to the Collaborative Adaptive Rangeland Management experiment is unclear.

> **Key Message 3:** Ecosystems across the Northern Great Plains provide recreational opportunities and other valuable goods and services that are at risk in a changing climate. Rising temperatures have already resulted in shorter snow seasons, lower summer streamflows, and higher stream temperatures, and have negatively affected high-elevation ecosystems and riparian areas, with important consequences for local economies that depend on winter or river-based recreational activities. Climate-induced land-use changes

in agriculture can have cascading effects on closely entwined natural ecosystems, such as wetlands, and the diverse species and recreational amenities they support.

Key Message 3 concerns the ecological and economic consequences of rising temperatures. The three pathways by which climate change affects recreation are clear, but winter recreation is not included in the list. The data supporting the economic importance of these pathways are weak and citations should be specific to this region. The paragraph on whitebark pine (page 930, line 37 to page 931, line 5) has no associated economic analysis and seems out of place with the other examples. It could be moved to the draft NCA4 Chapter 7, "Ecosystems, Ecosystem Services, and Biodiversity," or at least referenced in that chapter. In the case of low-elevation areas with a matrix of natural and agricultural lands, at least a note should be added stating that it is difficult to attribute loss of natural habitat to climate change, if no studies have quantified them. Attribution of climate change is important given the emphasis of the key message to use past trends as evidence of how the communities and industries will or will not adapt to future climate changes.

Key Message 3 does not consider the iconic national parks (e.g., Yellowstone, Glacier, Badlands, Grand Teton) and monuments (Charles Russell, Teddy Roosevelt, Devils Tower, Mount Rushmore) in the region and the importance of these protected regions for tourism and recreation. Likewise, the impact of climate change on managed and protected ecosystems (endangered species, fire, invasive species, wildlife transmitted diseases), as well as the socioeconomic challenges facing gateway communities, should be discussed. Headwaters Economics offers several important studies that could be considered by the chapter authors.[9]

> **Key Message 4:** Fossil fuel and renewable energy production and distribution infrastructure is expanding within the Northern Great Plains. Climate change and extreme weather events put this infrastructure at risk, as well as the supply of energy it contributes to support individuals, communities, and the U.S. economy as a whole. The energy sector is also a significant source of greenhouse gases and volatile organic compounds that contribute to climate change and ground-level ozone pollution.

Key Message 4 is fairly speculative and weakly supported by the provided references. A large percentage of the statements derive from U.S. Department of Energy reports, which are not very detailed or referenced on the points about infrastructure and transport that are emphasized in the key message. Whereas some of the impact pathways listed in this section are fairly logical and plausible, it is not easy to discern them as measurable impacts on energy supply or prices relative to the huge extraneous forcings on the energy sector from global supply, demand, and geopolitics. Economic modelling studies that could be cited to boost the case made here would strengthen the statement.

In terms of greenhouse gas mitigation, the chapter might mention the relative importance of coal-fired power plants in this region. The U.S. Environmental Protection Agency data suggest that Wyoming and North Dakota have the highest per capita emissions in the country. The authors should refer to the draft NCA4 mitigation (Chapter 29) and energy sector (Chapter 4) chapters, including Department of Energy initiatives related to carbon capture and sequestration.

---

[9] See https://headwaterseconomics.org/economic-development/trends-performance/montanas-economy-and-protected-lands and https://headwaterseconomics.org/dataviz/national-park-service-units.

> **Key Message 5:** Indigenous peoples of the Northern Great Plains are at high risk from a variety of climate change impacts, especially those resulting from hydrological changes (e.g., including changes in snowpack, glacier melt, seasonality and timing of precipitation events, extreme flooding and droughts, and reduction in streamflows). These changes are already resulting in harmful impacts to tribal economies, livelihoods, and sacred waters and plants used for ceremonies, medicine, and subsistence. At the same time, many tribes have been very proactive in adaptation and strategic climate change planning.

Key Message 5 includes good supporting evidence for anticipated and current impacts to culturally-important resources and the livelihoods of Indigenous Peoples. However, it is important to clarify which examples and citations come from the Northern Great Plains region. The invasive species case study and paragraph about proactive adaption add useful context.

Inclusion of Chapter 22 Materials in Chapter 1, "Overview"

The chapter's key messages are broadly covered in the draft NCA4 report findings and Chapter 1, "Overview." Inclusion of the Northern Great Plains in Chapter 1 could be expanded to better reflect issues of concern in the region. For example, on page 43, lines 32-34, declines in snowpack and shifts to more precipitation falling as rain is a clear component of climate projections for the Northern Great Plains and mountains of Montana and Wyoming, but this region is not identified in the discussion of this topic.

Comments on Graphics

Figure 22.1 should show the regional patterns and changes in annual and seasonal temperature and precipitation so that the reader has easy reference to this information at the regional scale. Referring back to draft NCA4 Chapter 2 national-scale Figure 2.6 is not convenient for this regional information. See "Other Recommended Changes" section for expanded discussion related to this figure recommendation.

In Figure 22.6, it is unclear what the map is supposed to be showing. More explanation is needed so that it can be understood as a stand-alone image.

Comments on Literature Cited

The chapter provides many recent and helpful citations from the peer-reviewed scientific literature. The Committee suggests that the authors consider other relevant literature (see Appendix B in this review report).

Comments on Traceable Accounts

The traceable accounts section is quite short—but provides some evidence that the findings are documented in a transparent and credible way. It is noted that each chapter in the draft NCA4 handles traceable accounts differently, with some, like this chapter, providing new information not previously discussed in the chapter and being relatively short. The Committee suggests better alignment of key message text and the traceable accounts to avoid seeming like afterthoughts, and where possible, more inclusion of specific examples of adaptation with the key messages.

Key Message 1 is assigned "very high confidence," but this does not align well with the text that follows (page 946), which notes uncertainties in changes in precipitation and run-off and precipitation variability. Moreover, the key message as written focuses on water management, whereas the analysis of uncertainty focuses on climate projections. The "Description of confidence and likelihood" states high confidence for warming temperatures, which is not specifically part of Key Message 1.

For Key Message 2, the points discussed in "Major uncertainties" and "Description of confidence" refer to climate change metrics but not the key message, which addresses producer responses. To call out more specifically the current adaptation measures that are underway by a "subset of producers" to support Key Message 2 would greatly strengthen the message and evidence base.

Key Message 3 text does not discuss "future government policies that could exacerbate or mitigate climate-induced losses," so this point seems poorly supported. Similarly, Key Message 4 does not discuss renewable energy, biofuel production, or low-level ozone production in any substantial way, although they are emphasized in the "Major uncertainties" section.

Other Recommended Changes

The opening "Background" section of Chapter 22 would benefit from a more detailed geographic description. For example, reference is made to three distinct geographic features—Red River Valley, Upper Missouri River Basin, mountains of Montana and Wyoming—but other subregions are not identified (e.g., Sand Hills, High Plains, North Platte River basin, etc.) and some are cited in the text but not shown on a map (e.g., the Snake River drainage, the Prairie Potholes region, and the Columbia River drainage). It would be helpful to have a map that shows the three identified geographic features, as well as a description of the other parts of the Northern Great Plains not covered by these three features. Figure 22.1 should show the regional patterns and changes in annual and seasonal temperature and precipitation so that the reader has easy reference to this information at the regional scale.

Table 22.3 does not include Montana prairie pothole data and it is unclear why this is the case.

More specific line comments for this chapter are provided in Appendix B.

CHAPTER 23: SOUTHERN GREAT PLAINS

Summary

Chapter 23 surveys the diverse set of anticipated climate change impacts on the industries, communities and ecosystems of the Southern Great Plains region. The chapter maintains an appropriate balance of attention across these issues by emphasizing what are likely to be the most sensitive and costly impacts, including those related to Gulf Coast infrastructure and industry, regional agricultural systems, and human health. The chapter is very readable for a general audience, making good use of discussions that telescope from big picture context down to engaging local case studies, including several from high-profile extreme events experienced in the region in recent years. The chapter does a good job of reflecting current scientific understanding via references to relevant research, including peer-reviewed literature and gray literature reports. It also links directly and clearly to findings of the CSSR and helpfully incorporates estimations of the economic costs of regional climate change impacts.

Suggested improvements to Chapter 23 are mostly around clarification and contextualization. For example, better descriptions are needed of baselines and how climate changes are expected to affect deviations from those baselines. Tables, figures and case study boxes should be better used to reinforce key messages. The Chapter 23 authors are also encouraged to ensure that there is appropriately balanced representation of locations across the region.

Comments Related to the Statement of Task

Comments on Key Messages

Key messages are not comprehensive, but generally provide well-chosen key takeaways for the chapter.

> **Key Message 1:** The region's growing population, the migration of individuals from rural to urban locations, and climate change will increase and redistribute demand and result in resource contention at the intersection of food consumption, energy production, and water resources. This "nexus" is inextricably linked to quality of life, particularly in rural areas as well as across both national and transnational borders.

In Key Message 1, the authors should reconsider whether the "nexus" language is necessary. It will come across as jargon to most general-audience readers. The main point is that disparate water users from different sectors, including food and energy production sectors, experience increased tension and tradeoffs during periods of water scarcity. Some further interaction effects are referenced but are not well-developed (e.g., water needed for electricity generation, which is needed for irrigation, which is needed for food production). Unless the Chapter 23 authors think that readers need to view these potentially interacting effects in a sophisticated way, the "nexus" term does not seem appropriate and it would suffice to point out that these industries are interdependent on each other and potential competitors for scarce water resources. Also, the Committee suggests considering adding language to Key Message 1 that clearly distinguishes it as a paragraph that is specific to the Southern Great Plains region. As currently worded, it could potentially be applicable to several U.S. regions. Adding a few words of geographic specificity would make it more engaging.

> **Key Message 2:** Higher temperatures, extreme precipitation, and rising sea levels associated with climate change make the built environment in the Southern Plains increasingly vulnerable to disruption, particularly as infrastructure ages and populations shift to urban centers. Coastal infrastructure remains particularly at risk as most climate projections suggest sea level rise of up to four feet if emissions are not reduced.

In Key Message 2, authors should consider adding a timeline for the statement on "up to four feet" sea level change. It would also be useful to make the sea level rise impacts in this key message specific to the Southern Great Plains region, since sea level rise impacts may be broadly applicable to U.S. coastlines.

Comments on Graphics

Tables, figures, and case study boxes should be better aligned with key messages. The case study boxes for Key Messages 1 and 2 and the specific examples for Key Message 3 and 4 are helpful. Key Message 5 should include a case study.

Figures 23.1 and 23.2 come across as overly-specific and somewhat arbitrary.

A map or infographic that show geographically the patterns and features of the region described in the "Background" section of the chapter would also be useful.

Comments on Literature Cited

Generally, the chapter reflects the peer-reviewed literature on the covered topics. It would be beneficial to provide a discussion of how the state of knowledge on regional climate change impacts has advanced since the NCA3, to provide better continuity.

Comments on Traceable Accounts

In contrast to some other chapters, the traceable accounts section is not used to provide an adequate set of references. The "Description of evidence base" explains only the general source of information for a particular key message, whereas most chapters provide specific evidence. The Chapter 23 authors should make better use of the traceable accounts section for strengthening supporting evidence for key messages, increasing transparency, and providing resources for readers wishing to explore any particular topic in the chapter in more detail.

The justifications for assignments of likelihood and confidence levels are not always clear and could be strengthened with additional references and comments on how the literature supports the claim. Descriptions of confidence and likelihood for Key Message 1 (page 992) and Key Message 2 (page 993) are the main examples where increased treatment is needed, but all justifications could be strengthened.

Other Recommended Changes

For readers not already intimately familiar with the geography of the Southern Great Plains, the "Background" section (first five paragraphs) of the chapter helps set baselines and provide basic context. The Chapter 23 authors should consider expanding this part of the chapter to include background descriptions of some of the other attributes of the region that are referred to in the key messages, such as tribal communities, natural habitats, agriculture, and population centers. In other words, make sure that background paragraphs are well-aligned with key messages.

The Committee suggests showing region-specific climate projections or anomalies (temperature, precipitation, days over 90°F, drought) where possible. The draft NCA4 Figure 2.6 shows projections for the entire United States, but it would be good to have a closer view of this region for easier reading of the linkages between climate projections and the impacts discussed in the chapter.

Local details provided in the chapter are heavily geared toward Texas. It is recommended that the material either be rebalanced somewhat, or explicitly state the reasoning for this approach. Is the emphasis on Texas a reflection of the relative states-of-knowledge across the region, or distributions of populations or economy size? The main relevant adaptation efforts of each state should be mentioned. For example, the Texas State Water Plan is noted, but efforts in Oklahoma or Kansas are not. Similarly, the discussion about managing risk with respect to ecosystems discusses programs for Texas only.

Message consistency with the draft NCA4 Chapter 10, "Agriculture and Rural Communities," should be reviewed and edits made accordingly. Chapter 23 currently includes at least one impact pathway that is not addressed in Chapter 10—grain quality—and conversely Chapter 10 includes some impact pathways that are relevant to the Southern Great Plains region that are not touched on here.

Also see line-by-line comments in Appendix B.

CHAPTER 24: NORTHWEST

Summary

Overall, the Committee found Chapter 24 to be strong and effective in communicating climate impacts, risks, and response actions in the Northwest. The framing of vulnerable communities as being on the front lines of climate change is excellent and should be considered as a framing approach for other chapters of the draft NCA4. The chapter draws appropriately on recent literature, provides clear and consistent messaging, and highlights examples well in included boxes. Graphics are clear and effective. More balanced discussion of species distribution changes and expanded discussion of demonstrated benefits of climate adaptation and mitigation actions in the region would improve this chapter.

Comments Related to the Statement of Task

Comments on Key Messages

Organization and content of the key messages is clear, consistent, and makes the findings accessible. The information is linked to observed climate and regional risks, future climate relevant to regional risks, the challenges, response actions, and success stories for reducing risk, and emerging issues. The language is accessible and effective at making climate impacts and possible actions to address them tangible and relatable to readers.

> **Key Message 1:** Climate change is already affecting the Northwest's diverse natural resources, which support sustainable livelihoods and provide a robust foundation for Tribal and rural communities. Increasing temperatures, changing precipitation patterns, and changes in coastal ocean waters have already reduced agricultural and fishery

> productivity, while also providing new business opportunities for parts of the natural resource economy. Climate change is expected to continue affecting the natural resource sector, valued at over $180 billion per year, but the economic consequences will depend on future market dynamics and adaptation efforts. Proactive management can increase the resilience of natural resources and economies.

The supporting text for Key Message 1 needs to provide evidence for the statement that "new business opportunities…" are provided for parts of the economy. Additionally, the Chapter 24 authors should point out potential effects of management and variability in adaptive capacity (i.e., "what can humans do about this?"). Evaluation of fishery management under climate change is cited in the draft NCA4 Chapter 9, "Oceans and Marine Resources," and mentioned in this chapter. Proactive management will help in some areas or sectors, but is harder to realize in others. This can be documented, even generally if no regional examples exist. The mention on page 1018 of the draft chapter needs citation.

> **Key Message 2:** Valued aspects of Northwest heritage and quality of life—the natural environment, wildlife, outdoor recreation, and Tribal cultures—will change with the climate. Increasing temperatures, reduced water availability, changing snow conditions, forest fires, habitat fragmentation, and other changes are endangering the well-being of a wide range of wildlife, threatening popular recreational activities and tribal subsistence and culture. For the Tribes, the health and vitality of the salmon runs is a direct indicator of the wider health of the region.

Key Message 2 should include a tie-in to observed consequences from extreme climate events, and what they indicate for the future. This is done in the text on page 1013.

> **Key Message 3:** Existing water, transportation, and energy infrastructure already face challenges from flooding, landslides, drought, wildfire, and heat waves. Future climate change raises the risk for many of these extreme events, potentially compromising the reliability of water supplies, hydropower, and transportation across the region. Isolated communities and those with systems that lack redundancy are the most vulnerable. Adaptation strategies that address more than one sector, or are coupled with social and environmental co-benefits, can increase resilience.

> **Key Message 4:** The ability of regional social and healthcare systems to expand quickly beyond normal service levels will fall short if cascading or acute hazards occur, exacerbating existing socioeconomic disparities. In addition to an increased likelihood of acute hazards and epidemics, disruptions in local economies and food systems could result in more chronic health risks. Organizations and volunteers that make up the Northwest's collective safety net are already stretched thin with current demands and will be further challenged by climate stressors. The potential health co-benefits of future climate mitigation investments could help to counterbalance these risks.

> **Key Message 5:** Communities on the front lines of climate change experience the first, and often the worst, effects. Frontline communities in the Northwest include Tribal and Indigenous peoples, the economically disadvantaged, and those most dependent on natural resources for their livelihoods. These communities generally prioritize basic needs, such as shelter, food, and transportation; frequently lack economic and political

capital; and have fewer resources to prepare for and cope with climate disruptions. However, the social and cultural cohesion inherent in many of these communities provides a foundation for building community capacity and increasing resilience.

In Key Messages 3, 4, and 5, the framing of vulnerable communities as being on the front lines of climate change is excellent. It is a way to draw in diverse communities who are most likely to be affected, which is not just based on socioeconomics, but also on livelihood dependency on ecosystems, location, etc. The Committee recommends this language be used elsewhere in the NCA4, as it helps greatly in orientating readers to the direct relevance of the key messages to them, or to communities they know. See also "Other Recommended Changes" section of this chapter review for additional suggestions related to Key Message 3.

Comments on Literature Cited

The chapter does a good job of reflecting current understanding and literature and is systematic in the treatment of climate science issues raised in the draft NCA4 Chapter 2, "Our Changing Planet," and how these issues affect the Northwest region. Furthermore, the report provides robust documentation of findings since the publication of the NCA3, although explicitly noting what is new could be improved.

Comments on Traceable Accounts

The "Description of evidence base," "Major uncertainties," and "Description of confidence and likelihood" sections do a good job for each key message in documenting statements, their uncertainties, and confidence associated with summaries. Highlighting new information provided since the NCA3 would help emphasize how knowledge about and confidence in climate impacts and adaptation and mitigation strategies has evolved in recent years.

Other Recommended Changes

More minor concerns that should be addressed include:

- Examples provided in the chapter imply that shifts in species distributions will all be detrimental to human communities and economies, which is not always the case. For example, in commercial fishery impacts introduced on page 1017, species are entering or will likely enter new regions, becoming available for catch where they were not previously. For tribes and others with accustomed fishing grounds, there will be winners and losers. Cheung et al. (2015) is cited later in the chapter, which is appropriate, but the point should be acknowledged consistently. Citations included in the draft NCA4 Chapter 9 also contain examples (e.g., Ianelli et al., 2001; Seung and Ianelli, 2016).
- The Northwest is one of a few regions in the United States that is strongly influenced by interannual and interdecadal climate variability associated with the Pacific Ocean. It would be useful to refer to Box 2.1 (in the draft NCA4 Chapter 2) on climate variability and discuss specific challenges of detecting climate impacts in the midst of the large climate variability in the Northwest, and what challenges this presents for developing adaptation strategies.

- The chapter should provide expanded discussion of the interactions among processes which are well documented in the region. For example, fire, water quality, and hydropower generation are inter-connected phenomena and sectors. This chapter could also link to more specific treatments of those individual topics in other relevant NCA4 chapters.
- It is suggested that the authors consider more clearly articulating that many of the "success" stories for climate adaptation are technological in nature. This provides an opportunity to also point out use of ecosystems as part of solutions, and many co-benefits that can be provided. For example, in the text describing Key Message 3 (infrastructure), the threat is well discussed, but the authors should also include in the "solutions" discussion the opportunities to use hybrid green and gray infrastructure to reduce flood and erosion risk. See The Nature Conservancy in Washington "Floodplains by Design"[10] work and citations provided in the draft NCA4 Chapters 8, "Coastal Effects," and 25, "Southwest."
- Mitigation gets little treatment this chapter, except in Key Message 4. There is available literature to cite that can be found in the draft NCA4 Chapter 8, "Coastal Effects," Chapter 9, "Ocean and Marine Resources," and possibly others, on integrative, positive effects of habitat and ecosystem protection and restoration on mitigating carbon emissions, in addition to the other adaptation co-benefits that such habitat-based strategies provide. See page 1026 for an example of where this could be discussed.
- Cross-border issues with Canada could receive more attention in this chapter. A straightforward way to do this could be to cite relevant information in the draft NCA4 Chapter 16, "Climate Effects on U.S. International Interests."
- The chapter could refer readers to state and local climate assessments and climate plans.
- It would be helpful if urban climate issues were given more emphasis.

CHAPTER 25: SOUTHWEST

Summary

This regional chapter is informative, up to date, and well written. It builds from the Southwest Chapter in the NCA3 and presents essential climate-related issues under seven key messages and selected highlight boxes. Traceable accounts effectively describe the evidence base and uncertainty of the key messages. The chapter gives excellent treatment to a full range of biophysical, socioeconomic and cultural impacts and provides strong examples of adaptation and mitigation strategies that give the reader tangible cases to increase understanding.

The Committee recommends more consistent discussion across the main text and traceable accounts sections of the chapter and increased citation of relevant literature for select topics. Some topics would benefit from increased treatment in the chapter, including coastal habitats and their role in protecting shorelines, urban climate issues, and interactions among processes that are well documented in the region such as fire, flooding, sedimentation, and cost of hydropower generation.

---

[10] See http://www.washingtonnature.org/floodplains.

Comments Related to the Statement of Task

Comments on Key Messages

> **Key Message 1:** Water supplies for people and nature in the Southwest are decreasing during droughts due in part to human-caused climate change. Intensifying droughts, increasingly heavy downpours, and reduced snowpack are combining with increasing water demands from a growing population, aging infrastructure, and groundwater depletion to reduce the future reliability of water supplies.

Key Message 1 neglects to mention flooding. Extreme events (droughts and floods) are more likely under future climate. It is fair to highlight drought in this key message, but flooding should not be neglected in the narrative of the main chapter body. Currently, flooding is mentioned in the traceable accounts section, but it is buried in the final "Evidence" section, making the documentation and key message narrative inconsistent. Besides an increase in the number of dry days, occasional wetter very wet days are expected to occur (Das et al., 2013; Polade et al., 2015, 2017). This key message could also draw linkage between shorter- and longer-term dryness—more dry days in projected future climate leads to more dry years (see Berg and Hall, 2015; Polade et al., 2017). Alternatively, Key Message 1 could be framed around extreme events, saying mostly these will be manifest as drought, but as evidence base discussion elaborates, depending on assumptions in models, flooding can sometimes occur, as recent observed events support (Odigie and Warrick, 2017).

Supporting text for Key Message 1 states that, "Models project substantial changes in snowpack, which supplies almost all of the water in the region…" (page 2091, lines 34-35). Stating, "almost all" is an exaggeration. Considerable runoff and stream discharge in California and other Southwest states is from rainfall. The fraction of water supplied varies considerably over the landscape, so it would be more appropriate to say something similar to "snowpack, which supplies a major portion of the water used in the region."

Supporting text for Key Message 1 would also benefit from discussion of desalinization as an alternative water supply and the high cost and environmental impacts of this process (see Cooley and Phurisamban, 2016).

> **Key Message 2:** The integrity of Southwest forests and other ecosystems and their ability to provide natural habitat, clean water, and economic livelihoods have declined as a result of recent droughts and wildfire due in part to human-caused climate change. Carbon emissions reductions, fire management, and other actions can help address future vulnerabilities of ecosystems and human well-being.

Key Message 2 could note that the seasonal occurrence and variability of downslope winds in California and elsewhere are usually accompanied by anomalously low humidity (Guzman-Morales et al., 2016).

> **Key Message 3:** Homes, beaches, fish, and other coastal resources in the Southwest have experienced sea level rise, ocean heating, ocean acidification, and reduced oxygen, all manifestations of human-caused climate change. Coastal infrastructure, marine plants and wildlife, and people who depend on fishing confront increased risks under continued climate change.

The supporting text for Key Message 3 should acknowledge the solid work done by California on protecting and restoring coastal habitats, and the resilience benefits they provide for erosion, flood reduction, nursery habitat, etc. The text could also discuss climate change impacts on physical, chemical, and biological components of the coast and estuaries (e.g., multiple facets of the Santa Barbara Coastal Ecosystem Vulnerability Assessment; Myers et al., 2017) and a cascade of possible impacts in the San Francisco Bay, where altered regional patterns of temperature, precipitation and sea level could cascade to provoke local impacts such as modified water supplies, increasing risks of coastal flooding, and growing challenges to sustainability of native species (see Cloern et al., 2011).

Some expanded information on the implications of variability in ocean acidification for ecosystems and people would be useful in the supporting text for this key message. Also, in the traceable accounts "Major uncertainties" it is unclear why there is no mention of variation in ocean acidification levels and vulnerability of species, similar to what is provided for sea level rise, especially since it is mentioned in the main body of the chapter.

> **Key Message 4:** Traditional foods, livelihoods, cultural resources, and spiritual well-being of Indigenous peoples in the Southwest are affected by drought, wildfire, and ocean warming. Because future changes could disrupt the ecosystems on which Indigenous peoples depend, tribes are developing adaptation measures and emissions reduction actions.

Key Message 4 presents little to no documentation for how livelihoods will be affected, despite highlighting it in the message. In the traceable accounts "Description of evidence base" (page 1118), no citations directly estimating such effects are provided. Livelihood diversification is notoriously difficult to estimate, and the chapter authors do not cite literature showing whether vulnerabilities in livelihoods occur, or if substitutes can buffer communities from nature-based economic changes. References should be cited, or acknowledgement provided that the connection is inferential, with discussion qualified as uncertain or a current research gap.

> **Key Message 5:** Renewable hydropower in the Southwest has shown declines during drought, due in part to climate change. Continued temperature increases, energy use from a growing population, and water competition with farms and cities reduce the future reliability of fossil fuels and hydropower. Renewable solar and wind energy are increasing and offer future options to cut carbon emissions and reduce water use.

Key Message 5 might benefit from mention of daily to multi-year variation in coastal cloud cover, which would affect solar electricity generation along the California coast. See Appendix B for specific references that could be cited for this topic.

> **Key Message 6:** Availability of food and viability of rural livelihoods are vulnerable to water shortages in the Southwest. Increased drought and reduction of winter chill can harm crops and livestock, exacerbate competition for water among food production, energy generation, and residential uses, and increase future vulnerabilities of food security and rural livelihoods.

For Key Message 6, uncertainty around the connection between climate impacts and both food security and rural livelihoods should be revised. It would be more accurate to include "and *may* increase future vulnerabilities of food security and rural livelihoods."

Similar to the comment on livelihoods made for Key Message 4, discussing "rural livelihoods" in Key Message 6 as increasingly vulnerable is not well supported by documentation in the narrative following it. Are there estimates of number of farmers, laborers, and other input providers, or transport and processing jobs associated with changes in agricultural production? How much of agricultural production in the region is from small holder producers versus larger industrial operations possibly more buffered from climate impacts?

> **Key Message 7:** Heat-associated deaths and illnesses, vulnerabilities to disease, and other health risks to people in the Southwest increase in extreme heat and in climate conditions that foster the growth and spread of pathogens. Improving stressed public health systems, community infrastructure, and personal health can reduce serious health risks under future climate change.

Some additional recent literature that should be cited to support Key Message 7 is listed in Appendix B comments for this chapter.

Comments on Graphics

Overall, Chapter 25 makes good use of simple visualizations and examples in figures and boxes.

For Figure 25.8, it is suggested that the time period of projected changes be inserted into the figure caption, e.g., "Projected increases in extremely hot days, (2036-2065 versus 1976-2005)."

Comments on Literature Cited

The chapter provides systematic and robust treatment of the topics discussed in the draft NCA4 Chapter 2, "Our Changing Climate" and appropriately discusses these topics in the context of the Southwest region. Recent literature produced since publication of the NCA3 is well cited throughout the chapter, with additional references suggested in relevant locations throughout this chapter review.

Comments on Traceable Accounts

Overall, the key messages are documented in a consistent, transparent, and credible way. However, some information is missing or inconsistently treated among main chapter text and traceable accounts sections. Specific comments on this are provided in the "Comments on Key Messages" section for this chapter review.

Other Recommended Changes

The chapter would benefit from expanded discussion of a few additional topics relevant to the region.

- Little treatment is given to coastal habitats (dunes, wetlands, seagrasses, etc.) and their role in protecting shorelines from erosion and flooding. As sea levels rise and king tide flood events increase, actions of California Coastal Commission and others in protecting coastal habitats are an important adaptation strategy.

- Increased discussion of interactions among processes in the region such as fire, flooding, sedimentation and cost of hydropower generation would be valuable (see Writer et al., 2014; Sankey et al., 2017 for examples).
- Mention of riverine and coastal flooding interactions is also suggested (Bromirski and Flick, 2008; Cayan et al., 2008), with an example of how longer-period climate changes and short-period weather extremes that arise from both oceanic and atmospheric processes may conspire in creating large impacts (such as in the San Francisco Bay/Delta).
- More attention could be given to United States-Mexico cross-border issues relevant to the region (see GNEB, 2016), such as drought, flood and wildfire occurrence (Westerling et al., 2004).
- The Chapter 25 authors should make it clear that "availability of food" is a general vulnerability due to potential declines in production of specialty crops, as stated on page 1122, and not a vulnerability specific to the Southwest region, which is not dependent on locally grown food for subsistence.
- Page 1097, lines 16-18 in the draft NCA4 state "Under the highest emissions scenario, climate change could raise sea level at San Francisco 30 in (76 cm) ± 11 in (28 cm) by 2100 (Griggs et al. 2017) to a maximum of 6.2 ft (1.9 m) (Jevrejeva et al. 2016)." The maximum stated here is misleading: the highest sea level rise reported in Griggs et al. (2017) approaches 3 meters.
- Page 1089, lines 32-34 in the draft NCA4 state "Hotter temperatures have already contributed to a 20% reduction of snowpack and its water content since 1950 (Pierce et al. 2008; Fyfe et al. 2017), with more than half of this attributed to human-caused climate change (Pierce et al. 2008; Pederson et al. 2011; Fyfe et al. 2017)." The Committee generally agrees with this statement, but 20% is debatable.
- Urban climate issues, including changes and barriers to adaptation, might be given more emphasis in the chapter (see Auffhammer and Aroonruengsawat, 2011; Ekstrom and Moser, 2014; Sun et al., 2015; Vahmani et al., 2016).
- A number of web-accessible items could also be considered. For readers interested in adaptation and planning, Chapter 25 might point to regional and community level efforts and climate data resources that have been conducted or are ongoing. Websites and other examples for these topics are listed in Appendix B.
- Readers could also benefit from reference to available state and local climate assessments and climate plans resources. Reference to other relevant draft NCA4 chapters should also be added, including those focused on water, oceans, and ecosystem services.

CHAPTER 26: ALASKA

Summary

This chapter clearly describes the effects of climate change in Alaska and is accurate, up-to-date, and rigorous. It provides substantial detail on the nature and geographic variation in climate change impacts. It also points to the recent and likely future changes but is cautious not to claim more than can be said from the existing data in both the main text and traceable accounts. Like other regional chapters, Chapter 26 is relatively detailed and perhaps more technical than some chapters. This makes it particularly helpful to Alaska-focused policy-makers, managers, and the public, but perhaps less accessible to some readers from other regions. For the general reader, additional details, such as a definition of permafrost, would be helpful.

One suggestion from the Committee is that the chapter be more explicit about the types of adaptation measures being implemented and the extent to which these currently meet adaptation needs. Much of the adaptation section is about the tools that are available and it provides fewer specifics than other parts of the chapter.

Comments Related to the Statement of Task

Comments on Key Messages

The key messages accurately reflect the understanding described in the text of the chapter with respect to past and projected changes and the associated risks for ecosystems and society. The key messages convey the main topics of the chapter, but are vaguer than the main text and traceable accounts.

> **Key Message 1:** Retreating and thinning arctic summer sea ice plays an important role on Alaska's marine wildlife and fish habitats, distributions, and food webs, all of which are important to Alaska's residents. These changes are anticipated to continue with unabated increases in CO2 emissions, which will accelerate ecosystem alterations that are difficult to predict.

The statement in Key Message 1 that retreating sea ice plays an important role in marine habitats seems somewhat vague; what exactly are the impacts? In the second sentence, is the intended message that future carbon dioxide emissions are expected to continue unabated? Or does it mean that *if* emissions continue unabated, these effects are expected to happen? The use of more concise and explicit language is recommended. Consider also whether the time interval (e.g., next 25 years or next 100 years) is important to specify.

Key Message 2: Local Alaskan residents, communities, and their infrastructure continue to be affected by permafrost thaw, coastal and river erosion, increasing wildfire, and glacier melt. These changes are expected to continue into the future with increasing warming temperatures, which will directly impact how and where many Alaskans will live.

Climate change impacts on infrastructure are described in Key Messages 2-5. Perhaps these could be consolidated in Key Message 5, which is explicitly about infrastructure damage. For example, in Key Message 3, why is damage to infrastructure listed as a human health risk in the same sense as injuries, smoke inhalation, and infectious disease?

> **Key Message 3:** Climate change brings a wide range of human health threats to Alaskans including increased injuries, smoke inhalation, damage to vital infrastructure, decreased food and water security, and new infectious diseases. The risks are greatest for rural residents who face physical harm from storms and flooding, loss of vital food sources, disrupted traditional practices, and who must consider relocation. Further adaptation

strategies would reduce the physical, social, and psychological harm likely to occur under a warming climate.

In the last sentence in Key Message 3, is it not the *implementation* of adaptation (rather than the strategies) that would reduce harm? Clarification is needed.

> **Key Message 5:** Climate warming is causing damage to infrastructure that will be costly to repair or replace, especially in remote Alaska. It is also reducing heating costs throughout the state. These effects are likely to grow with continued increases in greenhouse gas emissions. Timely repair and maintenance of infrastructure can reduce the damages and avoid some of these added costs.

For Key Message 5, "effects are likely to grow with continued increases in emissions." Is this sentence intending to state that there *will be* continued increases in emissions? See also comment for Key Message 1.

> **Key Message 6:** Proactive adaptation in Alaska can reduce costs, generate social and economic opportunity, and improve livelihood security. Direct engagement and partnership with communities is a vital element of adaptation in Alaska.

In Key Message 6, the first sentence states that adaptation can reduce costs and have other positive outcomes. However, it is unclear whether this is a statement of belief about consequences of adaptation in general or is a conclusion based on evaluation of adaptation actions that have been taken in Alaska; the traceable accounts indicate high confidence. The supporting text related to Key Message 6 states that adaptation actions are under way but provides little information on what adaptation actions and consequences have been observed. The supporting text should be expanded to provide concrete examples to support the text and confidence given.

Comments on Graphics

Most figures show changes in the physical environment (climate and erosion), not impacts on ecosystems or people (except for reduced heating costs), whereas the key messages emphasize effects on ecosystems and communities. Graphics should be selected to more clearly illustrate the key messages.

Comments on Literature Cited

The chapter does an excellent job of synthesizing the key literature, especially the literature since the NCA3 was published. The chapter makes one important misinterpretation of the literature though, which should be corrected. It states that climate change is causing conversion of forest to shrubland. The cited study (Mann et al., 2012) describes a change from conifer forests to deciduous vegetation (such as the aspen mixed-wood of Alberta). The statement that forests are changing to shrublands may be true, but the Mann et al. reference does not make this claim.

Comments on Traceable Accounts

The findings are documented in a consistent, transparent, and credible way, with traceable accounts supporting the key messages and likelihood evaluated appropriately, but see comments on key messages.

Other Recommended Changes

The chapter describes how adaptation could be done and what tools are available, but it does not describe explicitly what adaptation implementation is occurring. Some documentation of adaptation actions that have been implemented would be helpful.

See Appendix B for specific line comments for Chapter 26.

CHAPTER 27: HAWAI'I AND PACIFIC ISLANDS

Summary

This chapter on the assessment of climate change impacts and responses seen in Hawai'i and U.S. Affiliated Pacific Islands is an informative narrative on the conditions faced and addressed by a diverse group of island communities in the Pacific. It builds from the Hawai'i and U.S. Affiliated Pacific Islands Chapter in the NCA3 and presents essential climate-related issues. The assessment is consistent in acknowledging the diversity in situations while connecting the area through localized and regional data and information. The literature citations cover timelines and topics that were not addressed by the NCA3 for this region and include the progress made since the NCA3 in a useful manner. It is noteworthy that in the introduction the authors specifically identify the new areas of research and data relevant to the draft NCA4 and this region. The chapter is readable and provides notable indicators of inclusivity in a complicated process of information gathering in a clear and admirable manner. The use of case studies and strategic figures (including the map of the region, Figure 27.1) strengthens the chapter, and there is a high level of attention paid to including adaptation efforts or comments throughout. The introduction clearly describes the uniqueness of the region and the way climate issues pose challenges, the linkages to other U.S. islands and the fact that there is uncertainty regarding some projections and impacts, while acknowledging that there is also action being taken through policy and adaptation initiatives. The information and narrative under each of the six key messages supports the points being offered. The chapter is consistent in the level of effort exerted to have the reader understand the uncertainty and risk being experienced by the Pacific Island communities as a consequence of the changing climate. Traceable accounts effectively describe the process and evidence base of the key messages.

Comments Related to the Statement of Task

Comments on Key Messages

Key Message 1: Dependable and safe water supplies for Pacific Island communities and ecosystems are threatened by rising temperatures, sea level rise, and increased risk of extreme drought and flooding. Islands, especially low atolls, already experience saltwater contamination due to sea level rise, which could catastrophically impact food and water security. Active monitoring and management of watersheds and freshwater systems could increase resilience to future threats.

Key Message 1 refers to results from different forms of statistical and dynamical downscaling. These methods may not yield consistent outcomes since statistical downscaling assumes stationarity of underlying processes and dynamical downscaling less so. Dynamic downscaling may be better suited to understanding Pacific Island climate changes, especially precipitation changes, because it can potentially better represent fine scale topographic influences. See also the *Other Recommended Changes* section of this chapter review.

Key Message 4: Fisheries and the livelihoods they support are threatened by warmer ocean temperatures and ocean acidification. Widespread coral reef bleaching and mortality have recently occurred in successive years, and by midcentury these events are projected to occur annually. Bleaching and acidification will result in loss of reef structure, leading to lower fisheries yields and loss of coastal protection and habitat. Declines in oceanic fishery productivity of up to 15%, and 50% of current levels are projected by midcentury and 2100, respectively.

The supporting text for Key Message 4 could be clarified. For instance, the text states that, "If the current emissions trajectory continues, coral reefs will experience annual bleaching beginning in about 2035 in the Marianas Archipelago, in about 2040 in American Sāmoa and the Hawaiian Islands, and in about 2045 at other equatorial reefs" (page 1257, lines 5-7). Many readers may not equate an emissions trajectory to a given amount of warming. Also, the amount of warming projected during the next few decades is roughly the same for all emissions scenarios. This statement could be modified to, "under projected warming of approximately 0.5°F per decade, coral reefs will experience…" or something similar.

Comments on Graphics and Boxes

For Box 27.1, additional context should be provided for the statement "El Niño-Southern Oscillation (ENSO) is the main source of year-to-year climate variability in Hawai'i and USAPI." ENSO does have a strong influence on interannual variability, but sea level fluctuations have also been linked quite strongly to multi-year variability, such as the Pacific Decadal Oscillation and decadal trade wind strengthening and weakening (Bromirski et al., 2011; Merrifield, 2011; Hamlington et al., 2016). These lower frequency fluctuations and links to sea level (and probably rain and snowfall) should be discussed. Additionally, rainfall has been decreasing in Hawai'i, presumably because of changes in trade wind regime or North Pacific storm tracks. It would be informative to readers to know what mechanisms are involved and what season(s) have experienced the greatest changes. In general, in addition to reporting projected local changes a description of causal mechanism provides useful insight.

Figure 27.7 may mislead readers because it is presented in rates of sea level change (inches per decade), in comparing a short period (2012-2015) with a longer period (1993-2011). Granted, the rates have changed dramatically, but the overall amount of anomalous sea level rise in 1993-2011 is large and that in 2012-2015 is relatively small. One possible way to address this could be to instead show the time average sea level anomaly for the two periods.

For Figure 27.9, "The Marshallese Traditional Agroforestry Calendar," the fact that such a calendar exists and may be used is interesting, but the Committee doubts that most readers will be able to decipher this diagram. It would also take a lot of mental effort to distinguish the El Niño version from the traditional version; showing one of these versions is more than enough.

Comments on Literature Cited

The literature cited reflects both literature that demonstrates the advancement of data and information availability in climate science in the region, as well as relatively recent research studies that were published before 2013 addressing topics specific to island issues that do not have recent updates. In general, presentation is linked to findings of the CSSR and its summary in the draft NCA4 Chapter 2.

Comments on Traceable Accounts

Overall, the key messages are documented in a consistent, transparent and credible way. In the "Description of confidence and likelihood" section for Key Message 3, it is stated that: "There is very high confidence that a continued rise in global temperature will lead to increases in the rate of sea level rise." The Committee agrees with this, but it is suggested that the authors add that while there is high confidence that sea level will rise, there is not much confidence in the actual amount of sea level rise—the increased rate could fall within a wide range of possible outcomes.

The process that was used to prepare this chapter seems to have effectively enlisted scientists (Workshop) and Stakeholders (Town Hall meetings), and the evidence base is well documented by reference to numerous sources.

Comments on Data and Analyses

In general, there is consistency in the way that data and analyses are handled. But for the most part, these are not new analyses. It should be noted that on page 1257, line 25, the work cited from Bell et al. (2013) reports a 20% decline in coral reef fish production under model projections from an SRES A2 emissions scenario (still relevant, but rather dated) in contrast to the RCP approach being used more commonly throughout the draft NCA4 report.

Other Recommended Changes

- As an overview, it would be useful to have U.S. Island systems on both sides of the globe (Hawai'i Pacific Island, this chapter, and the U.S. Caribbean, Chapter 20) affirm the uniqueness and similarities that islands face in addressing climate change issues. This chapter handled the Pacific side, but in a self-contained fashion. It is important that linkages or comparisons to the U.S. Caribbean chapter of the draft NCA4 be made, with cross-referencing between the two included, as appropriate. An issue that both might address is whether the temperature and precipitation changes projected for these island systems is to large extent the same as changes projected over the broader sweep of oceanic area within which they are placed, or whether the island landscape and topography creates important differences from those larger-scale open ocean changes.

- The "Summary Overview" section of this chapter addresses the topics and information at a reasonable level; however, the last paragraph that begins on page 1233 (lines 35-38) and ends on page 1234 (lines 1-5) might benefit from additional attention to flow and connectivity of ideas and information. Chapter 27 is longer than some of the other chapters and a case could be made for reducing the number of pages, but the Committee would not recommend it if doing so would mean a loss of the sense of process and an understanding of the complexity of the area and its challenges.
- In the reporting of temperature (e.g., page 1243, line 15 and page 1271, lines 6-7), the authors report an air temperature increased by 0.42°C (0.76°F) in the past 100 years. This is surely an estimate from a limited set of weather stations, which probably have some degree of uncertainty. The way it is stated could be interpreted as being highly confident with a two decimal place accuracy. Care should be taken here and elsewhere to ensure that statements do not falsely exude high confidence.
- Page 1247, lines 3-5 states that, "Throughout the region, the number of climate and water resources monitoring stations has declined (Oki, 2004; Keener et al., 2012; Giambelluca et al., 2013), reducing the ability of researchers to project future changes in climate." This might also be inserted in the traceable accounts section for Key Message 1, since it introduces uncertainty (or more certainty if observations are shored up).

CHAPTER 28: NEAR-TERM ADAPTATION NEEDS AND INCREASED RESILIENCY

Summary

This chapter on adaptation response is one of the stronger chapters in the draft NCA4 report. It is balanced, easy to read, and frames the discussion well in terms of a challenge that can be tackled by investing and adapting. The tables and figures are effective and relatable for conveying important messages for the intended audience. As noted elsewhere in this review report, expanded and integrated treatment of climate change responses across the draft NCA4 is strongly encouraged, including discussion and examples of adaptation. Chapter 28 provides a robust foundation for that discussion. The risk management conversation is timely and positive. This is an important message for the intended audiences—adaptation is a form of risk management. The key messages are effective and accurate, but could be improved by incorporating stronger language that ties in concrete examples.

Comments Related to the Statement of Task

Comments on Key Messages

The key messages reflect current understanding and current and emerging conversations, especially in terms of risk management and risk reduction. They are clear and consistent, and reflect a balance of challenges and options for responding. However, there are stronger statements in the body of the chapter that could be woven into the key messages to strengthen their impact.

Page 1309, lines 36-39 states, "Mainstreaming of climate adaptation into existing decision processes has begun in many areas, for instance in financial risk reporting, capital investment planning, engineering standards, military planning and disaster risk management." This sentence is strong and effective in that is uses concrete examples.

Page 1311, line 20 states, "Adaptation is a form of risk management." This is a simple, yet strong sentence. It puts adaptation in familiar risk management terms. Intended audiences can relate to this as a manageable challenge when framed this way.

Comments on Graphics and Tables

Figure 28.1 is a simple yet effective graphic and illustrates action and progress well. Any organization could scale and use this information.

Table 28.1 is also a simple, effective, and relatable tool for intended audiences, including decision-makers.

Comments on Literature Cited

Overall, Chapter 28 accurately reflects the peer-reviewed literature and appears to illustrate progress since the NCA3 was published. The chapter should rely on peer-reviewed literature, gray literature, or direct source data from cities, states or agencies that meets the NCA4 quality standards. It is suggested that the authors avoid quoting newspapers (e.g., page 1318 cites the *Miami Herald*) that may not fully capture a policy, project, or situation.

Comments on Traceable Accounts

Chapter findings are documented in a consistent, transparent and credible manner with high to medium levels of confidence. The findings are communicated effectively and rather easily for intended audiences.

Other Recommended Changes

The Southeast Florida Regional Climate Change Compact is a model of regional climate adaptation planning.[11] Recent developments include the creation of the 2017 updated Regional Climate Action Plan.[12] This activity would be a strong example to emphasize in this chapter. The Miami Beach stormwater infrastructure program is also a good example which demonstrates progress from regional planning to local adaptation action.[13] Miami Beach is often noted as a city investing and adapting and not just planning. It is recommended that the NCA4 authors review the work of the Compact and the City of Miami Beach for possible reference and inclusion.

---

[11] See http://www.southeastfloridaclimatecompact.org.
[12] See http://www.southeastfloridaclimatecompact.org.
[13] See http://miamibeachfl.gov/risingabove.

Many other adaptation efforts at local scales have been established across the U.S., which could be drawn on for examples in the draft NCA4, including the Bay Area Regional Collaborative,[14] the King County-Cities Climate Collaboration,[15] the Metropolitan Mayors Coalition,[16] and the Climate Compact of Colorado Communities,[17] among others.

On page 1321, line 16, the statement about fuel treatments reducing risks of wildfire should be qualified by adding "in some forests."

CHAPTER 29: MITIGATION: AVOIDING AND REDUCING LONG-TERM RISKS

Summary

Mitigation is aimed at reducing the magnitude of human-caused climate change, which complements the use of adaptation to cope with climate change impacts. The draft chapter provides an overview of how mitigation efforts can affect climate change risk and impacts. It makes the appropriate points that (1) adaptation and mitigation are complementary efforts; (2) the scale of risks and impacts can be reduced through mitigation; and (3) the timing and magnitude of emission reductions are important for reducing risk.

In the "State of Mitigation" section of this draft chapter, too much focus is placed on the U.S. government's withdrawal from the Paris Agreement. A large number of mitigation responses have been undertaken beyond the federal government—by individuals, local and state governments, large corporations, and many other institutions—and these efforts are insufficiently covered in this chapter. An expanded inclusion of important examples of a range of mitigation initiatives would offer a solutions-oriented message, a perspective that would align more clearly with the broader messaging approach recommended in this review report.

The Chapter 29 authors are encouraged to describe the timescales associated with mitigation versus adaptation, namely that mitigation provides benefits on the longer timescales of interest (approximately 100 years) and probably has a limited effect in the near term (approximately 25 years). Placing focus on the timing of mitigation is necessary to highlight the important point that without significant emission reductions, the United States will experience substantial and far-reaching impacts, especially in the latter half of the century. This could be emphasized and discussed more explicitly in the chapter.

Chapter 29 focuses largely on economic values and essentially argues from a cost-benefit framing. The Committee sees the value in including economic risks and impacts in the draft NCA4, which are primarily only included in this chapter. As noted in the comments in the "Front Matter: Report Findings" section earlier in Chapter 3 of this review, the revised NCA4 should address economic impacts more broadly in the national topic and regional chapters to complement the discussion of impacts. That said, the draft Chapter 29 focuses too narrowly on economic impacts, and thus does not reflect the risk-based framing of the rest of the report.

---

[14] See http://bayarearegionalcollaborative.org/projects.html.
[15] See https://kingcounty.gov/services/environment/climate/strategies/k4c.aspx.
[16] See https://www.mapc.org/our-work/expertise/climate/mmc.
[17] See https://www.compactofcoloradocommunities.org.

Review Comments Related to the Statement of Task

Comments on Key Messages

For the most part, the chapter's key messages are clear, but they are incomplete and not anchored in a risk framing around likelihood and consequence. Messages could also be framed in a more solution-oriented manner to make it more accessible to broad audiences. Generally, it is recommended that the messages place greater emphasis on response actions than on potential impacts.

It may also be worthwhile to include a key message that emphasizes that mitigation operates on time scales that span multiple decades that may have mid-course iteration while investments in adaptation can expect more immediate or short-term benefits.

> **Key Message 1:** Recent scientific advances in impact quantification demonstrate that climate change under a high emissions scenario and without adaptation will impose substantial physical and economic damages on the United States economy, human health, and the environment, with the potential for annual losses in some sectors reaching hundreds of billions of dollars by the end of the century. Some impacts, such as sea level rise from ice sheet disintegration, will be irreversible for thousands of years, while others, such as species extinction, will be permanent.

Key Message 1 focuses on the risk of inaction, imprecisely applied. The message would be clearer if it started with "Without significant mitigation efforts, impacts on the United States are expected to be substantial," rather than starting with impact quantification.

> **Key Message 2:** Substantial global-scale greenhouse gas emissions reductions are shown to significantly reduce climate change impacts and economic damages across the United States, though the magnitude and timing of avoided risks varies by sector and region.

Key Message 2 would be improved by revising to "the magnitude and timing of avoided risks varies by sector, region *and population adaptive capacity (or socioeconomics*)." The second paragraph in the support text (page 1355, lines 20-29) supports this additional caveat.

> **Key Message 3:** Adaptation can complement mitigation due to already committed climate change from past and present emissions and the inability to avoid all climate risks. Adaptation can reduce exposure and vulnerability to climate change in the United States in a variety of sectors. Recent studies have made advancements in capturing complex interactions between mitigation and adaptation including both benefits and adverse consequences.

The Committee suggests a thorough rethinking the content of Key Message 3. The first two sentences only convey information on the well-known complementarity of adaptation and mitigation. The last sentence about recent studies simply attributes high confidence that progress has been made in recent research.

Comments on Treatment of Risk

The focus on economic valuation in Chapter 29 does not reflect the risk management perspective of the rest of the report. The chapter acknowledges that adaptation efforts will reduce impacts, but it does not directly estimate or account for how these efforts will reduce impacts and net costs. This is understandable though, because data to support calibrating the value of adaptation on an aggregate level is lacking. The selected approach essentially argues from a cost-benefit framing and only calibrates risks and benefits in dollars. Adopting more calibrations of risk, and recognizing the importance of timing of mitigation action, would make this chapter much more powerful, up to date, and broadly consistent with the treatment of risk in other chapters of the draft NCA4. To be more specific, economic damages are only one part of the equation. The chapter reports aggregate economic damages without recognizing that even the most current literature is incomplete in coverage and inadequate in reflecting adaptation. Some mention of this fundamental caveat is needed and would make the discussion more consistent with recent literature on this topic (Hsiang et al., 2017).

Inclusion of additional approaches to quantifying aggregate economic damages would strengthen the chapter. It is recommended that the "Reasons for Concern" framework (Schneider et al., 2007; O'Neill et al., 2017) be used. This framework is used to communicate risks associated with climate change based on scientific evidence and expert judgement and would move the chapter content beyond the economics. Aggregate damages under multiple emissions scenarios can be estimated in decadal increments driven by transient temperature change (see Yohe, *Climatic Change*, November 2017; Hsiang et al., 2017; O'Neill et al., 2017). This work includes uncertainty in emissions and associated temperature change for each temperature target, and emphasizes dependence of damages and concerns on observed transient temperature change. More broadly, considerable research has been published recently for inclusion in the forthcoming IPCC Special Report on 1.5°C and should be cited in this chapter, which could support and inform Key Messages 2 and 3.

Comments on Graphics and Tables

Figure 29.1 is effective and a welcome addition to the draft NCA4. See specific comments provided in the review of the draft NCA4 Chapter 1, "Overview," found earlier in Chapter 3 of this review report, where this figure is included as Figure 1.5.

Figure 29.2 is interesting and effective at communicating the relative ranking of impacts from mitigation, but confidence in the estimates is overstated by not giving ranges and reporting too many significant figures. The figure reports estimates of year 2090 damages along RCP8.5 running from hundreds of billions of dollars per sector down to 1 million. That reflects six significant figures in accuracy from top to bottom, which is not credible. This should be clearly explained.

Figure 29.3 is interesting, but will be challenging for many readers. It should be revised to better highlight the differences across the three RCP scenarios and link them to mitigation, if possible.

Table 29.1 is an annotated bibliography of studies which conveys no content and is not comprehensive. This table could be useful if it provided information on what the noted studies concluded.

Comments on Literature Cited

As noted elsewhere in this chapter review, there are some gaps in recent literature on risk management, mitigation values beyond economic measures, time frame differentiation, and recognition of the iterative nature of responding to climate change.

Section 29.5.2, "Reducing Risk Through Climate Intervention," is appropriate for this chapter. The conclusion about carbon dioxide removal, however, should be modified. It is incomplete in stating only that carbon dioxide removal is "estimated to have high costs and long implementation times...." It should also mention that several studies of long term climate mitigation strategies suggest that without carbon capture and storage, the long-term costs of addressing climate change are much higher. The IPCC Fifth Assessment Synthesis Report (IPCC, 2014) concluded that carbon capture and storage technology will be essential to meet more near-term climate goals, such as the mid-century climate goal of keeping global temperature rise within 2°C agreed to in the Paris Agreement. In fact, IPCC suggests that without carbon capture, utilization, and storage, mitigation costs will rise by 138%. Finally, the Chapter 29 authors might want to mention that considerable effort is underway to reduce the cost of carbon capture, utilization, and storage, including finding alternative uses of the captured carbon, (e.g., the Carbon XPrize[18]).

The chapter could also better link to topic chapters in some instances, such as the discussion of agricultural carbon sequestration could also note that this would improve soil water retention (page 1358, line 23).

Comments on Traceable Accounts

The findings are consistent and transparent with what has been presented. However, there is a larger body of available literature that could be draw on to support the chapter text, which the Committee recommends the Chapter 29 authors consider including to bolster the traceable accounts.

Comments on Data and Analyses

The data and analyses included in this chapter are consistent, transparent, and credible, to a point. The assessment is limited and narrow because it does not critically evaluate the data presented. Critical evaluation of weaknesses, strengths, omissions, and other caveats reported in the cited literature should be noted.

Other Recommended Changes

In the draft chapter discussion of aggregate economic damages, the text should be clear that reported totals are dominated by health impacts derived from the value of statistical life (not a well-accepted concept, though better within a country than across the globe).

Readers may be confused when the topics switch between impacts and mitigation. For example, Section 29.4, jumps from the previous section on emission reductions to impacts

---

[18] See https://carbon.xprize.org/teams.

quantification. To help with this transition, it would be useful for the opening paragraph to mention this connection. Specifically, "to understand how mitigation can help reduce impacts, it is useful to look at how the impacts change under various emission scenarios." In addition, headers that always mention mitigation or make the connection between mitigation and impacts would reduce confusion. Organizing in terms of likelihood (confidence) and consequence (including adaptation in the reported literature) would also be useful.

While the ancillary benefits of mitigation are mentioned in Section 29.5.1, this section is relatively short and could be expanded to explain that the immediate benefits are also ones that tend to be especially beneficial for vulnerable populations and tend to have large public support.

Given that it is challenging to determine how adaptation will impact the net cost of climate change calibrated in dollars, human lives, likelihoods of extreme weather events, the distribution of impacts, and the potential of crossing irreversible tipping points, the chapter provides only the beginning of an adequate overview of how mitigation can reduce long-term climate risk. While expanded discussion of the synergy between adaptation costs and mitigation is likely infeasible in the NCA4, it would be helpful to provide a brief overview of this topic. As stated in the IPCC Synthesis Report (2007), "Responding to climate change involves an iterative risk management process that includes both adaptation and mitigation and takes into account climate change damages, co-benefits, sustainability, equity, and attitudes to risk."

This chapter could benefit from framing existing and needed mitigation and adaptation efforts (and their interactions) in an adaptive iterative risk-management framework, with cross-reference to the draft NCA4 Chapter 17, Sectoral Interdependencies, Multiple Stressors, and Complex Systems. Reference to other national topic chapters where mitigation is possible is also suggested (e.g., Chapter 7, "Ecosystems, Ecosystem Services, and Biodiversity," Chapter 10, "Agriculture and Rural Communities," etc.).

References

Auffhammer, M., and A. Aroonruengsawat (2011), Simulating the impacts of climate change, prices and population on California's residential electricity consumption, *Climatic Change*, *109*(SUPPL. 1), 191-210, doi:10.1007/s10584-011-0299-y.

Ayyub, B. M., Risk Analysis in Engineering and Economics, Chapman and Hall/CRC Press, second edition, 2014.

Ayyub, B. M., and Klir, G. J., Uncertainty Modeling and Analysis in Engineering and the Sciences, Chapman & Hall/CRC, 2006.

Ayyub, B. M., and Wright, R. N., 2016. "Adaptive Climate Risk Control of Sustainability and Resilience for Infrastructure Systems," Editorial, J Geography and Natural Disasters, 6(2), http://dx.doi.org/10.4172/2167-0587.1000e118.

Bakker, A. M. R., D. Louchard, and K. Keller (2017), Sources and implications of deep uncertainties surrounding sea-level projections, *Climatic Change*, *140*(3), 339-347, doi:10.1007/s10584-016-1864-1.

Balazs, C. L., R. Morello-Frosch, A. E. Hubbard, and I. Ray (2012), Environmental justice implications of arsenic contamination in California's San Joaquin Valley: A cross-sectional, cluster-design examining exposure and compliance in community drinking water systems, *Environmental Health: A Global Access Science Source*, *11*(1), doi:10.1186/1476-069X-11-84.

Barnett, K., S. A. Parks, C. Miller, and H. T. Naughton (2016), Beyond fuel treatment effectiveness: Characterizing interactions between fire and treatments in the US, *Forests*, *7*(10), doi:10.3390/f7100237.

Bautista, E., J. C. Osorio, and N. Dwyer (2015), Building climate justice and reducing industrial waterfront vulnerability, *Social Research*, *82*(3), 821-838.

Benz, J., T. Tompson, and J. Agiesta (2014), The People's Agenda: America's Priorities and Outlook for 2014*Rep.*, 17 pp, The Associated Press-NORC.

Berg, A., J. Sheffield, and P. C. D. Milly (2017), Divergent surface and total soil moisture projections under global warming, *Geophysical Research Letters*, *44*(1), 236-244, doi:10.1002/2016GL071921.

Berg, N., and A. Hall (2015), Increased interannual precipitation extremes over California under climate change, *Journal of Climate*, *28*(16), 6324-6334, doi:10.1175/JCLI-D-14-00624.1.

Bradley, C. M., C. T. Hanson, and D. A. DellaSala (2016), Does increased forest protection correspond to higher fire severity in frequent-fire forests of the western United States?, *Ecosphere*, *7*(10), doi:10.1002/ecs2.1492.

Bravo, M. A., R. Anthopolos, M. L. Bell, and M. L. Miranda (2016), Racial isolation and exposure to airborne particulate matter and ozone in understudied US populations: Environmental justice applications of downscaled numerical model output, *Environment International*, *92-93*, 247-255, doi:10.1016/j.envint.2016.04.008.

Bromirski, P. D., and R. E. Flick (2008), Storm surge in the San Francisco Bay/Delta and nearby coastal locations, *Shore & Beach 76*.

Bromirski, P. D., A. J. Miller, R. E. Flick, and G. Auad (2011), Dynamical suppression of sea level rise along the Pacific coast of North America: Indications for imminent acceleration, *Journal of Geophysical Research: Oceans*, *116*(7), doi:10.1029/2010JC006759.

Brooks, J. S., T. M. Waring, M. Borgerhoff Mulder, and P. J. Richerson (2018), Applying cultural evolution to sustainability challenges: an introduction to the special issue, *Sustainability Science*, *13*(1), 1-8, doi:10.1007/s11625-017-0516-3.

Brown, J. F., and M. S. Pervez (2014), Merging remote sensing data and national agricultural statistics to model change in irrigated agriculture, *Agricultural Systems*, *127*, 28-40, doi:10.1016/j.agsy.2014.01.004.

Brulle, R. J., and D. N. Pellow (2006), Environmental justice: Human health and environmental inequalities, in *Annual Review of Public Health*, edited, pp. 103-124, doi:10.1146/annurev.publhealth.27.021405.102124.

Burke, M., and K. Emerick (2016), Adaptation to climate change: Evidence from US agriculture, *American Economic Journal: Economic Policy*, *8*(3), 106-140, doi:10.1257/pol.20130025.

Butler, B. W., R. D. Ottmar, T. S. Rupp, R. Jandt, E. Miller, K. Howard, R. Schmoll, S. Theisen, R. E. Vihnanek, and D. Jimenez (2013), Quantifying the effect of fuel reduction treatments on fire behavior in boreal forests, *Canadian Journal of Forest Research*, *43*(1), 97-102, doi:10.1139/cjfr-2012-0234.

Cayan, D. R., E. P. Maurer, M. D. Dettinger, M. Tyree, and K. Hayhoe (2008), Climate change scenarios for the California region, *Climatic Change*, *87*(1), 21-42, doi:10.1007/s10584-007-9377-6.

Chiara, N., J. Garvin Michael, and J. Vecer (2007), Valuing Simple Multiple-Exercise Real Options in Infrastructure Projects, *Journal of Infrastructure Systems*, *13*(2), 97-104, doi:10.1061/(ASCE)1076-0342(2007)13:2(97).

Clark, P. U., et al. (2016), Consequences of twenty-first-century policy for multi-millennial climate and sea-level change, *Nature Climate Change*, *6*(4), 360-369, doi:10.1038/nclimate2923.

Cloern, J. E., et al. (2011), Projected Evolution of California's San Francisco Bay-Delta-River System in a Century of Climate Change, *PLoS ONE*, *6*(9), e24465, doi:10.1371/journal.pone.0024465.

Cochrane, M. A., C. J. Moran, M. C. Wimberly, A. D. Baer, M. A. Finney, K. L. Beckendorf, J. Eidenshink, and Z. Zhu (2012), Estimation of wildfire size and risk changes due to fuels treatments, *International Journal of Wildland Fire*, *21*(4), 357-367, doi:10.1071/WF11079.

Cook, B. I., T. R. Ault, and J. E. Smerdon (2015), Unprecedented 21st century drought risk in the American Southwest and Central Plains, *Science Advances*, *1*(1), doi:10.1126/sciadv.1400082.

Cooley, H., and R. Phurisamban (2016), *The Cost of Alternative Water Supply and Efficiency Options in California*, Pacific Institute.

Das, T., E. P. Maurer, D. W. Pierce, M. D. Dettinger, and D. R. Cayan (2013), Increases in flood magnitudes in California under warming climates, *Journal of Hydrology*, *501*, 101-110, doi:10.1016/j.jhydrol.2013.07.042.

Douville, H., and M. Plazzotta (2017), Midlatitude Summer Drying: An Underestimated Threat in CMIP5 Models?, *Geophysical Research Letters*, *44*(19), 9967-9975, doi:10.1002/2017GL075353.

Ekstrom, J. A., and S. C. Moser (2014), Identifying and overcoming barriers in urban climate adaptation: Case study findings from the San Francisco Bay Area, California, USA, *Urban Climate*, *9*, 54-74, doi:https://doi.org/10.1016/j.uclim.2014.06.002.

Emanuel, K. A. (1999), Thermodynamic control of hurricane intensity, *Nature*, *401*(6754), 665-669, doi:10.1038/44326.

Feng, S., M. Oppenheimer, and W. Schlenker (2012), *Climate Change, Crop Yields, and Internal Migration in the United States*.

Fulé, P. Z., J. E. Crouse, J. P. Roccaforte, and E. L. Kalies (2012), Do thinning and/or burning treatments in western USA ponderosa or Jeffrey pine-dominated forests help restore natural fire behavior?, *Forest Ecology and Management*, *269*, 68-81, doi:https://doi.org/10.1016/j.foreco.2011.12.025.

Gao, J., et al. (2018), Public health co-benefits of greenhouse gas emissions reduction: A systematic review, *Science of the Total Environment*, *627*, 388-402, doi:10.1016/j.scitotenv.2018.01.193.

Gleick, P. H. (2014), Water, drought, climate change, and conflict in Syria, *Weather, Climate, and Society*, *6*(3), 331-340, doi:10.1175/WCAS-D-13-00059.1.

Good Neighbor Environmental Board, GNEB (2016), Climate Change and Resilient Communities Along the U.S.-Mexico Border: The Role of the Federal Agencies, edited.

Green, R., V. Sarovar, B. Malig, and R. Basu (2015), Association of stillbirth with ambient air pollution in a california cohort study, *American Journal of Epidemiology*, *181*(11), 874-882, doi:10.1093/aje/kwu460.

Guzman-Morales, J., A. Gershunov, J. Theiss, H. Li, and D. Cayan (2016), Santa Ana Winds of Southern California: Their climatology, extremes, and behavior spanning six and a half decades, *Geophysical Research Letters*, *43*(6), 2827-2834, doi:10.1002/2016GL067887.

Haimes, Y. Y. (2009). *Risk Modeling, Assessment, and Management*. Third edition. John Wiley & Sons, Hoboken, New Jersey.

Hall, J. W., R. J. Lempert, K. Keller, A. Hackbarth, C. Mijere, and D. J. McInerney (2012), Robust Climate Policies Under Uncertainty: A Comparison of Robust Decision Making and Info-Gap Methods, *Risk Analysis*, *32*(10), 1657-1672, doi:10.1111/j.1539-6924.2012.01802.x.

Hamlington, B. D., S. H. Cheon, P. R. Thompson, M. A. Merrifield, R. S. Nerem, R. R. Leben, and K. Y. Kim (2016), An ongoing shift in Pacific Ocean sea level, *Journal of Geophysical Research: Oceans*, *121*(7), 5084-5097, doi:10.1002/2016JC011815.

Hansen, A., L. Bi, A. Saniotis, and M. Nitschke (2013), Vulnerability to extreme heat and climate change: Is ethnicity a factor?, *Global Health Action*, *6*(1), doi:10.3402/gha.v6i0.21364.

Hatfield, J., G. Takle, R. Grotjahn, P. Holden, R. C. Izaurralde, T. Mader, E. Marshall, and D. Liverman (2014), Ch. 6: Agriculture. Climate Change Impacts in the United States: The Third National Climate Assessment, J. M. Melillo, Terese (T.C.) Richmond, and G. W. Yohe, Eds., U.S. Global Change Research Program, 150-174. doi:10.7930/J02Z13FR.

Herman, J. D., P. M. Reed, H. B. Zeff, and G. W. Characklis (2015), How should robustness be defined for water systems planning under change?, *Journal of Water Resources Planning and Management*, *141*(10), doi:10.1061/(ASCE)WR.1943-5452.0000509.

Hill, E. (2016). An Economic and Environmental Comparison of Solar and Fossil Fuel Energy. Retrieved from http://www.trunity.net/sam2/view/article/51cbf4837896bb431f6afc17.

Hsiang, S., R. Kopp, A. Jina, J. Rising, M. Delgado, S. Mohan, D. J. Rasmussen, R. Muir-Wood, P. Wilson, M. Oppenheimer, K. Larsen, and T. Houser (2017), Estimating economic damage from climate change in the United States, *Science*, *356*(6345), 1362-1369, doi:10.1126/science.aal4369.

Ianelli, J. N., A. B. Hollowed, A. C. Haynie, F. J. Mueter, and N. A. Bond (2011), Evaluating management strategies for eastern Bering Sea walleye pollock (Theragra chalcogramma) in a changing environment. ICES Journal of Marine Science 68:1297-1304.

Institute of Medicine (2007), *Environmental Public Health Impacts of Disasters: Hurricane Katrina: Workshop Summary*, 100 pp., The National Academies Press, Washington, DC, doi:10.17226/11840.

International Organization of Standardizations (ISO), 2009a. Risk Management—Principles and Guidelines, ISO 31000, iso.org, Geneva.

IPCC (2007), Climate Change 2007: Synthesis Report. Contribution of Working Groups I, II and III to the Fourth Assessment Report of the Intergovernmental Panel on Climate Change [Core Writing Team, Pachauri, R.K and Reisinger, A. (eds.)]. IPCC, Geneva, Switzerland, 104 pp.

IPCC (2012), Managing the Risks of Extreme Events and Disasters to Advance Climate Change Adaptation. A Special Report of Working Groups I and II of the Intergovernmental Panel on Climate Change [Field, C.B., V. Barros, T.F. Stocker, D. Qin, D.J. Dokken, K.L. Ebi, M.D. Mastrandrea, K.J. Mach, G.-K. Plattner, S.K. Allen, M. Tignor, and P.M. Midgley (eds.)]. Cambridge University Press, Cambridge, United Kingdom and New York, NY, USA, 582 pp.

IPCC (2014), Climate Change 2014: Synthesis Report. Contribution of Working Groups I, II and III to the Fifth Assessment Report of the Intergovernmental Panel on Climate Change [Core Writing Team, R.K. Pachauri and L.A. Meyer (eds.)]. IPCC, Geneva, Switzerland, 151 pp.

Jones, A. D., K. V. Calvin, W. D. Collins, and J. Edmonds (2015), Accounting for radiative forcing from albedo change in future global land-use scenarios, *Climatic Change*, *131*(4), 691-703, doi:10.1007/s10584-015-1411-5.

Kalies, E. L., and L. L. Yocom Kent (2016), Tamm Review: Are fuel treatments effective at achieving ecological and social objectives? A systematic review, *Forest Ecology and Management*, *375*, 84-95, doi:10.1016/j.foreco.2016.05.021.

Klir, G. J. (2005), *Uncertainty and Information: Foundations of Generalized Information Theory*, John Wiley & Sons, Inc., doi:10.1002/0471755575.

Kolb, C., M. Pozzi, C. Samaras, and M. VanBriesen Jeanne (2017), Climate Change Impacts on Bromide, Trihalomethane Formation, and Health Risks at Coastal Groundwater Utilities, *ASCE-ASME Journal of Risk and Uncertainty in Engineering Systems, Part A: Civil Engineering*, *3*(3), 04017006, doi:10.1061/AJRUA6.0000904.

Kronk-Warner, E. 2013. Tribal Renewable Energy Development Under the Hearth Act: An Independently Rational, But Collectively Deficient Option. Arizona Law Review, Vol. 55, No. 1031.

Lawrence, J., R. Bell, P. Blackett, S. Stephens, and S. Allan (2018), National guidance for adapting to coastal hazards and sea-level rise: Anticipating change, when and how to change pathway, *Environmental Science & Policy*, *82*, 100-107, doi:https://doi.org/10.1016/j.envsci.2018.01.012.

Lombardo, F. T., and Bilal, M.A (2015), Analysis of Washington, DC, Wind and Temperature Extremes with Examination of Climate Change for Engineering Applications, *ASCE-ASME Journal of Risk and Uncertainty in Engineering Systems, Part A: Civil Engineering*, *1*(1), 04014005, doi:10.1061/AJRUA6.0000812.

Mallakpour, I., and G. Villarini (2015), The changing nature of flooding across the central United States, *Nature Climate Change*, *5*(3), 250-254, doi:10.1038/nclimate2516.

Meisen, P., and T. Erberich Renewable Energy of Tribal Lands*Rep.*, 28 pp, Global Energy Network Institute.

Melillo, J. M., T. T. C. Richmond, and G. W. Yohe (2014), Climate Change Impacts in the United States: The Third National Climate Assessment*Rep.*, 841 pp, U.S. Global Change Research Program.

Merrifield, M. A., and M. E. Maltrud (2011), Regional sea level trends due to a Pacific trade wind intensification, *Geophysical Research Letters*, *38*(21), doi:10.1029/2011GL049576.

Millar, C. I., N. L. Stephenson, and S. L. Stephens (2007), Climate change and forests of the future: Managing in the face of uncertainty, *Ecological Applications*, *17*(8), 2145-2151, doi:10.1890/06-1715.1.

Milly, P. C. D., and K. A. Dunne (2016), Potential evapotranspiration and continental drying, *Nature Climate Change*, *6*(10), 946-949, doi:10.1038/nclimate3046.

Moftakhari, H. R., G. Salvadori, A. AghaKouchak, B. F. Sanders, and R. A. Matthew (2017), Compounding effects of sea level rise and fluvial flooding, *Proceedings of the National Academy of Sciences*, *114*(37), 9785-9790, doi:10.1073/pnas.1620325114.

Monahan, W. B., and N. A. Fisichelli (2014), Climate Exposure of US National Parks in a New Era of Change, *PloS One*, *9*(7), 13, doi:10.1371/journal.pone.0101302.

Moser, S. C., and L. Dilling (2011), Communicating Climate Change: Closing the Science-Action Gap, in *The Oxford Handbook of Climate Change and Society*, edited, pp. 161-176, Oxford University Press, Oxford, doi:10.1093/oxfordhb/9780199566600.003.0011.

Myers, M. R., Cayan, D. R., Iacobellis, S. F., Melack, J. M., Beighley, R. E., Barnard, P. L., Dugan, J. E. and Page, H. M., 2017. Santa Barbara Area Coastal Ecosystem Vulnerability Assessment. CASG-17-009.

NASEM (2016), *Characterizing Risk in Climate Change Assessments: Proceedings of a Workshop*, 100 pp., The National Academies Press, Washington, DC, doi: 10.17226/23569.

NASEM (2017a), *Communicating Science Effectively: A Research Agenda*, 152 pp., The National Academies Press, Washington, DC, doi:10.17226/23674.

NASEM (2017b), *Review of the Draft Climate Science Special Report*, 132 pp., The National Academies Press, Washington, DC, doi:10.17226/24712.

National Park System Advisory Board (2012), Revisiting Leopold: Resource Stewardship in the National Parks*Rep.*, 23 pp.

National Research Council (2013), *A Review of the Draft 2013 National Climate Assessment*, 130 pp., The National Academies Press, Washington, DC, doi:10.17226/18322.

National Research Council (2015), *Review of the Draft Interagency Report on the Impacts of Climate Change on Human Health in the United States*, 78 pp., The National Academies Press, Washington, DC, doi:10.17226/21787.

Odigie, K. O., and J. A. Warrick (2017), Coherence Between Coastal and River Flooding along the California Coast, *Journal of Coastal Research*, doi:10.2112/JCOASTRES-D-16-00226.1.

O'Gorman, P. A. (2014), Contrasting responses of mean and extreme snowfall to climate change, *Nature*, *512*(7515), 416-418, doi:10.1038/nature13625.

O'Neill, B. C., et al. (2017), IPCC reasons for concern regarding climate change risks, *Nature Climate Change*, *7*(1), 28-37, doi:10.1038/nclimate3179.

Oppenheimer, M., M. Campos, R. Warren, J. Birkmann, G. Luber, B.C. O'Neill, and K. Takahashi (2014), Emergent risks and key vulnerabilities. In: Climate Change 2014: Impacts, Adaptation, and Vulnerability. Part A: Global and Sectoral Aspects. Contribution of Working Group II to the Fifth Assessment Report of the Intergovernmental Panel on Climate Change [Field, C.B., V.R. Barros, D.J. Dokken, K.J. Mach, M.D. Mastrandrea, T.E. Bilir, M. Chatterjee, K.L. Ebi, Y.O. Estrada, R.C. Genova, B. Girma, E.S. Kissel, A.N. Levy, S. MacCracken, P.R. Mastrandrea, and L.L. White (eds.)]. Cambridge University Press, Cambridge, United Kingdom and New York, NY, USA, pp. 1039-1099.

Oppenheimer, M., C. M. Little, and R. M. Cooke (2016), Expert judgement and uncertainty quantification for climate change, *Nature Climate Change*, *6*, 445, doi:10.1038/nclimate2959. https://www.nature.com/articles/nclimate2959#supplementary-information.

Polade, S. D., A. Gershunov, D. R. Cayan, M. D. Dettinger, and D. W. Pierce (2017), Precipitation in a warming world: Assessing projected hydro-climate changes in California and other Mediterranean climate regions, *Scientific Reports*, *7*(1), doi:10.1038/s41598-017-11285-y.

Polade, S. D., D. W. Pierce, D. R. Cayan, A. Gershunov, and M. D. Dettinger (2014), The key role of dry days in changing regional climate and precipitation regimes, *Scientific Reports*, *4*, doi:10.1038/srep04364.

Prein, A. F., C. Liu, K. Ikeda, S. B. Trier, R. M. Rasmussen, G. J. Holland, and M. P. Clark (2017), Increased rainfall volume from future convective storms in the US, *Nature Climate Change*, *7*(12), 880-884, doi:10.1038/s41558-017-0007-7.

Prein, A. F., R. M. Rasmussen, K. Ikeda, C. Liu, M. P. Clark, and G. J. Holland (2016), The future intensification of hourly precipitation extremes, *Nature Climate Change*, *7*, 48, doi:10.1038/nclimate3168. https://www.nature.com/articles/nclimate3168#supplementary-information.

Ranger, N., T. Reeder, and J. Lowe (2013), Addressing 'deep' uncertainty over long-term climate in major infrastructure projects: four innovations of the Thames Estuary 2100 Project, *EURO Journal on Decision Processes*, *1*(3), 233-262, doi:10.1007/s40070-013-0014-5.

Rodriguez, I. (2017), Linking well-being with cultural revitalization for greater cognitive justice in conservation: Lessons from Venezuela in canaima national park, *Ecology and Society*, *22*(4), doi:10.5751/ES-09758-220424.

Sankey, J. B., J. Kreitler, T. J. Hawbaker, J. L. McVay, M. E. Miller, E. R. Mueller, N. M. Vaillant, S. E. Lowe, and T. T. Sankey (2017), Climate, wildfire, and erosion ensemble foretells more sediment in western USA watersheds, *Geophysical Research Letters*, *44*(17), 8884-8892, doi:10.1002/2017GL073979.

Schneider, S.H., S. Semenov, A. Patwardhan, I. Burton, C.H.D. Magadza, M. Oppenheimer, A.B. Pittock, A. Rahman, J.B. Smith, A. Suarez and F. Yamin, 2007: Assessing key vulnerabilities and the risk from climate change. Climate Change 2007: Impacts, Adaptation and Vulnerability. Contribution of Working Group II to the Fourth Assessment Report of the Intergovernmental Panel on Climate Change, M.L. Parry, O.F. Canziani, J.P. Palutikof, P.J. van der Linden and C.E. Hanson, Eds., Cambridge University Press, Cambridge, UK, 779-810.

Schwartz, R. E., A. Gershunov, S. F. Iacobellis, and D. R. Cayan (2014), North American west coast summer low cloudiness: Broadscale variability associated with sea surface temperature, *Geophysical Research Letters*, *41*(9), 3307-3314, doi:10.1002/2014GL059825.

Seung, C., and J. Ianelli. 2016. Regional economic impacts of climate change: a computable general equilibrium analysis for an Alaska fishery. Natural Resource Modeling 29:289-333.

Sherwood, S., and Q. Fu (2014), A drier future?, *Science*, *343*(6172), 737-739, doi:10.1126/science.1247620.

Sleeter, B. M., T. S. Wilson, E. Sharygin, and J. T. Sherba (2017), Future Scenarios of Land Change Based on Empirical Data and Demographic Trends, *Earth's Future*, doi:10.1002/2017EF000560.

Staudinger, M. D., N. B. Grimm, A. Staudt, S. L. Carter, F. S. Stuart Iii, P. Kareiva, M. Ruckelshaus, and B. A. Stein (2012), Impacts of climate change on biodiversity, ecosystems, and ecosystem services: technical input to the 2013 National Climate Assessment, *Report Rep.*, i-A-6 pp, Washington, D.C.

Steininger, K. W., B. Bednar-Friedl, H. Formayer, and M. König (2016), Consistent economic cross-sectoral climate change impact scenario analysis: Method and application to Austria, *Climate Services*, *1*, 39-52, doi:https://doi.org/10.1016/j.cliser.2016.02.003.

Sun, F. P., D. B. Walton, and A. Hall. 2015. A Hybrid Dynamical-Statistical Downscaling Technique. Part II: End-of-Century Warming Projections Predict a New Climate State in the Los Angeles Region. *Journal of Climate 28*, 4618-4636.

USGCRP (2016), *The Impacts of Climate Change on Human Health in the United States: A Scientific Assessment*, 312 pp., U.S. Global Change Research Program, Washington, DC, doi:10.7930/J0R49NQX.

USGCRP (2017), *Climate Science Special Report: Fourth National Climate Assessment, Volume I*, 470 pp., U.S. Global Change Research Program, Washington, DC, USA, doi: 10.7930/J0J964J6.

Vahmani, P., F. Sun, A. Hall, and G. Ban-Weiss. 2016. Investigating the climate impacts of urbanization and the potential for cool roofs to counter future climate change in Southern California. Environmental Research Letters 11.

Viscusi, W. K. (1994), Risk-Risk Analysis, *Journal of Risk and Uncertainty*, *8*(1), 5-17.

Walton, R. (2015), Ten Years After: How Entergy New Orleans survived Hurricane Katrina.

Wan, W., et al. (2017), Hydrological Drought in the Anthropocene: Impacts of Local Water Extraction and Reservoir Regulation in the U.S, *Journal of Geophysical Research: Atmospheres*, *122*(21), 11,313-11,328, doi:10.1002/2017JD026899.

Wanders, N., and Y. Wada (2015), Human and climate impacts on the 21st century hydrological drought, *Journal of Hydrology*, *526*, 208-220, doi:10.1016/j.jhydrol.2014.10.047.

Warren, J. L., J.-Y. Son, G. Pereira, B. P. Leaderer, and M. L. Bell (2017), Investigating the Impact of Maternal Residential Mobility on Identifying Critical Windows of Susceptibility to Ambient Air Pollution During Pregnancy, *American Journal of Epidemiology*, kwx335-kwx335, doi:10.1093/aje/kwx335.

West, J. J., S. J. Smith, R. A. Silva, V. Naik, Y. Zhang, Z. Adelman, M. M. Fry, S. Anenberg, L. W. Horowitz, and J.-F. Lamarque (2013), Co-benefits of mitigating global greenhouse gas emissions for future air quality and human health, *Nature Climate Change*, *3*, 885, doi:10.1038/nclimate2009 https://www.nature.com/articles/nclimate2009# supplementary-information.

Westerling, A. L., D. R. Cayan, T. J. Brown, B. L. Hall, and L. G. Riddle (2004), Climate, santa ana winds and autumn wildfires in southern california, *Eos*, *85*(31).

Wong, T. E., and K. Keller (2017), Deep Uncertainty Surrounding Coastal Flood Risk Projections: A Case Study for New Orleans, *Earth's Future*, *5*(10), 1015-1026, doi:10.1002/2017EF000607.

Woodward, M., Z. Kapelan, and B. Gouldby (2014), Adaptive Flood Risk Management Under Climate Change Uncertainty Using Real Options and Optimization, *Risk Analysis*, *34*(1), 75-92, doi:10.1111/risa.12088.

Wright, R. N., Ayyub, B. M., and Lombardo, F. T., "Bridging the Gap between Climate Change Science and Structural Engineering Practice," Structure Magazine of the National Council of Structural Engineers Associations (NCSEA), the Structural Engineering Institute (SEI) of ASCE, and the Council of American Structural Engineers (CASE) of the American Council of Engineering Companies (ACEC), 2013, 29-31.

Writer, J. H., A. Hohner, J. Oropeza, A. Schmidt, K. Cawley, and F. L. Rosario-Ortiz (2014), Water treatment implications after the High Park Wildfire in Colorado, *Journal-American Water Works Association*, *106*(4), 85-86, doi:10.5942/jawwa.2014.106.0055.

Ye, S., H. Y. Li, L. R. Leung, J. Guo, Q. Ran, Y. Demissie, and M. Sivapalan (2017), Understanding flood seasonality and its temporal shifts within the contiguous United States, *Journal of Hydrometeorology*, *18*(7), 1997-2009, doi:10.1175/JHM-D-16-0207.1.

Zhao, T., K. Sundararajan Satheesh, and C.-L. Tseng (2004), Highway Development Decision-Making under Uncertainty: A Real Options Approach, *Journal of Infrastructure Systems*, *10*(1), 23-32, doi:10.1061/(ASCE)1076-0342(2004)10:1(23).

Zoraster, R. M. (2010), Vulnerable populations: Hurricane Katrina as a case study, *Prehospital and Disaster Medicine*, *25*(1), 74-78, doi:10.1017/S1049023X00007718.

Appendix A. Comments on Appendix 5: Frequently Asked Questions

The Committee found the Frequently Asked Questions (FAQs) Appendix to be a useful resource and complement to the main text of the draft NCA4. The FAQs generally provide relevant information that is easily accessible to a non-technical audience.

For the FAQs that relate to specific draft NCA4 content, the provided information is generally consistent with the main text and it is useful when the main text is referenced. However, the emphasis and specific information sometimes differs between the FAQs and the main text. The FAQ information also seems out of date in some cases and should be updated to reflect the current state of knowledge.

The challenge of including a FAQs section is the inevitable feedback that there are other important and appropriate questions that could be addressed. For instance, there could be additional questions on islands and whether climate change impacts them differently, some of the other health-related challenges associated with a changing environment, and other topics. Draft NCA4 topics that do not get much treatment in the FAQs include water, agriculture, energy, ecosystems, and the interconnections between ecological and social systems. Additionally, there are no region-specific FAQs. There could also be an FAQ on scientific consensus. The included FAQs, "Why are scientists so certain that human activities are the primary cause of the recent global warming?" and "How reliable are the computer models of Earth's climate?" relate to important issues of scientific consensus, but do not address it directly. The lack of scientific consensus is a myth that could be addressed directly.

While a wide range of FAQs are relevant to the draft NCA4 and communicating climate science, the draft NCA4 authors could give further consideration to how FAQs are selected for inclusion. Are they intended to largely address climate science and the types of topics included in the CSSR report, or are they intended to be aligned with the content of the NCA report itself? Or both? Most of the FAQs focus on climate change science rather than impacts, risks, and responses that are the primary focus of the draft NCA4. Both types of information have value, but it may be useful to be more focused in what is included in Appendix 5 of the draft NCA4.

Technical review of each FAQ was provided by a committee member or a consultant with relevant expertise. Only FAQs where changes are recommended are listed in this section. Editorial comments for some FAQs are included in the line comments (see Appendix B). Regarding the inclusion of citations in the FAQs, it is recommended that they either be omitted or listed at the end of the FAQ, so as not to break up the text for the intended audience. As such, it is broadly suggested that all references cited in the FAQ answers be removed.

COMMENTS ON INDIVIDUAL FREQUENTLY ASKED QUESTIONS

How do we know Earth is warming?

Improved accuracy of this FAQ is needed. While some suggestions are provided here, it is recommended that a subject matter expert be consulted.

Early in the answer to this FAQ, it would be more accurate to note that earth is *generally* warming. The current text that there is warming could be misinterpreted and obscure the changes in extremes.

The statement, "Decades of temperature readings from thermometers and other scientific instruments around the world" (page 1444, lines 3-4) understates the robustness of the understanding of warming since the language arguably at least excludes data from satellites, boreholes, geological data, and other non-instrumental data sources. A more comprehensive statement is needed. Something similar to "multiple independent lines of evidence point conclusively to a warming planet. These include thermometer measurements…etc." would better reflect current understanding.

The phrase "volunteers or automated instruments" (page 1444, line 11) is a strange choice of language in that it excludes non-volunteer (i.e., professional) human observers, of which there must be some; if not now, then in the earlier years of the thermometer record. Also, mentioning volunteers may tend to devalue the reliability of the measurements. It would be better simply to say "At thousands of ground-based weather- and climate-stations around the world, instruments record."

Page 1444, lines 12-16 states, "Observations from these stations agree with readings from satellite instruments that measure land temperatures from space." Measuring land surface temperatures from satellites is extremely challenging, and reviewers are unaware of any datasets showing multidecadal global trends based on this approach. Would it be more appropriate to say "lower atmosphere" instead of "land"? The statement also refers to temperatures, not temperature trends, and thus may be strictly speaking true. However, readers could infer it to refer to temperature trends, and thus be misled. This should be clarified.

In the discussion of land ice in this FAQ, it is suggested that the Antarctic and Greenland ice sheets be specifically mentioned. It is recognized that these are sometimes included in the term "glaciers" but they merit specific, independent mention.

For Figure A5.2, the concept of "mass balance" in panel (f) may be too technical for the target audience without explanation. For panel (i) wildfire, the length of record shown is too short to establish a meaningful trend, and in fact there does not appear to be much of one. Lengthier comments on this figure are provided in response to the main text of the draft NCA4, where the graphic appears at Figure 1.1.

How is recent global warming different than warming in the past?

The supporting text for this FAQ implies that only warming epochs have been linked with orbital cycles, while attributing cooling epochs specifically to volcanic eruptions. It is the understanding of the reviewer that both warming and cooling epochs are forced by orbital cycles, with either a lack of volcanic eruptions or many volcanic eruptions playing primarily contributory roles to warming or cooling epochs. Orbital variations do not change the total amount of sunlight reaching the earth over the course of a year, but they affect its distribution by season and latitude, which then drive the growth or decay of high-latitude ice sheets and the strong positive climate feedback loops that go with them. The literature on this topic should be revisited by the draft FAQ authors to ensure the state of knowledge is accurately conveyed and the text should be revised to clarify the relationship between orbital cycles and both warming and cooling periods. Rewording of the supporting text (page 1447, lines 13-15) could state: "Warming *and cooling* epochs were *driven* by natural variations of the earth's orbit that altered the amount of sunlight that reached the Earth's Arctic and Antarctic regions, driving the retreat and advance of massive ice sheets. *Additionally, quiescent or active periods of volcanic eruptions also could contribute to warming or cooling epochs, respectively.*"

Page 1447, line 8, the phrase "at least" should be deleted. Long-term paleoclimatic time series from ocean sediments clearly show that the 100,000-year cycle extends back to roughly a million years ago; before that, a 40,000-year periodicity was dominant.

What's the difference between global warming and climate change?

The supporting text for this FAQ begins stating that "global warming" means a period when Earth's annual average surface temperature is increasing (page 1449, lines 12-13). Since global warming actually refers to the phenomenon of warming and not a time period, this statement is not technically accurate and the language should be revised.

The explanation of the difference between "climate change" and "global warming" currently states, "The entire globe isn't warming uniformly though" (page 1449, line 17). While this is true, this is not why the term "climate change" has a different meaning from "global warming." The main reason is, as explained elsewhere, the former term encompasses a wide range of phenomena in addition to an increase in global temperature. The sentence that follows (page 1449, lines 18-21) furthers this misinterpretation and it is recommended that it be deleted from the FAQ. The phrase "side effects" (page 1499, line 22) may seem trivializing. It is suggested something more specific, such as "associated consequences" be used. It is also unclear why this FAQ states that "climate change is less precise" (page 1450, line 6) and it is suggested that this statement be deleted.

Were there predictions of global cooling in the 1970s?

The primary point made in the italicized answer to this FAQ—that there is a preponderance of scientific literature noting warming occurred during this time period—seems to get lost, or even contradicted, in the supporting text. To approve readability and accuracy, it is suggested that the supporting text (beginning page 1450, line 25) be revised to something like "ending of ice ages led a few scientists in the 1970s to contemplate that the current warm interglacial period might be ending soon leading to a new ice age over the next few centuries. These few speculations were picked up and amplified by the media. But at that time there were far more scientific articles describing how warming would occur from the increase of greenhouse gases from the burning of fossil fuels (Figure A5.5). Scientists continue to study."

Also, for clarification, it is suggested that the last sentence on page 1450 (line 30) be revised to "the composition of the atmosphere in such a short period of time relative to natural orbital processes that the next ice age has now likely been delayed." The NCA4 authors could also consider noting that this delay could be tens of thousands of years (Clark et al., 2016; full citation in "References" section of this review report).

What are greenhouse gases?

Rather than (or perhaps in addition to) listing greenhouse gases on page 1451, lines 12-13, it is suggested that the text note that the buildup of greenhouse gases in Earth's atmosphere is the primary driver of recent warming and other climatic changes. It would also be useful to indicate that the anthropogenic greenhouse effect is an incremental increase in the existing (natural) greenhouse effect.

The supporting text could indicate that the molecular structure of greenhouse gases causes them to absorb terrestrial infrared radiation (unlike N_2 and O_2, the most abundant gases in the atmosphere).

Why are scientists so certain that human activities are the primary cause of recent global warming?

Important information is missing from this FAQ and it is suggested that additional experts be consulted to ensure the accuracy and completeness of the included materials. Additional topics that should be included are: (1) satellite measurements show no trend in energy output of the sun, and (2) paleo data show that warming observed since around 1900 is extremely unusual in the contact of the last 500-1,500 years.

On page 1453, lines 2-5, the provided argument suggests incorrectly that attribution of warming to human activities relies strongly on models. Paleo data show that there was actually a slow cooling trend in the centuries before the start of fossil fuel use, so this should be mentioned, with modeling studies discussed as a way to confirm that this observed cooling was due to orbital forcings.

The argument about the human origin of increased atmospheric greenhouse gases given on page 1453, lines 9-11, logically should be presented first, not last.

Citations that could be reviewed to support this FAQ include Marcott et al. (2013) and Marsicek et al. (2018).

How do we know that carbon dioxide is the main driver behind global warming?

The answer provided in response to this FAQ does not directly address the question. Focus on the recent rate of increase in atmospheric carbon dioxide does not in itself demonstrate that humans are the cause of the rise in temperature. Important arguments that should be noted here include the paleo record showing a very strong correlation between atmospheric carbon dioxide concentration and temperature going back hundreds of thousands of years. This would be a much more effective figure than what is currently illustrated in Figure A5.8.

Alternatively, the FAQ itself could be modified. Examples could be, "How do we know humans are changing the amount of carbon dioxide in the atmosphere? Or "How are humans changing the amount of carbon dioxide in the atmosphere?"

What role does water vapor play in global warming?

The explanation for this FAQ is difficult to understand and inaccurate in some places. On page 1456, lines 9-10 the FAQ response states, "water vapor cannot be the driver of global warming." One might argue that water vapor actually *is* a driver of global warming, since it is the dominant greenhouse gas. The point the FAQ authors may be trying to convey is that humans have no direct control over how much water vapor is in the atmosphere, although there is a strong indirect influence since evaporation and water vapor holding capacity increases with temperature. This should be stated more directly.

The language included on page 1456, lines 11-19 is likely too technical for the intended audience and it is recommended that it be deleted. In this text, line 12: "does accumulate" should probably be "does not accumulate" and line 12-13 "water vapor that would otherwise evaporate from the surface" is confusing because water vapor is already evaporated and direct versus indirect changes in water vapor will be difficult to understand for most readers. It might be better to say that water vapor is indeed the strongest greenhouse gas, with respect to current warming of the Earth's surface, but by itself it is not a strong driver of current and expected changes. Water vapor is not produced by combustion of fossil fuels, but it tends to be an amplifier of the warming effects caused by rising concentrations of carbon dioxide, methane, and nitrous oxide because a warmer atmosphere can hold more water vapor, thereby strengthening the associated greenhouse effect. See Myre et al. (2013) for quantification for the strength and sign of the water vapor feedback.

Figure A5.9. does not effectively communicate the main messages of this FAQ response and the inset graph is incomprehensible. It is recommended that this figure be deleted.

The response to this FAQ would be improved by simplifying it to begin with the second paragraph (page 1456, line 20) and also noting that humans cannot directly control the amount of water vapor in the atmosphere, although humans indirectly affect water vapor as global temperature rise from increasing greenhouse gas concentrations.

Have the sun or other natural factors contributed to the observed global warming of the past 50 years?

For page 1457, lines 10-11, it may be simpler to say directly that there is a cooling tendency from volcanic activity. According to the IPCC Fifth Assessment Report (IPCC, 2014), the estimated net effect of solar variability and volcanoes over this time period is essentially zero (but if anything a tiny warming tendency; IPCC Synthesis Report Figure 1.4, see also Chapter 2 of the draft NCA4). There may be a slight cooling tendency from orbital forcing, but over 50 years it would be minuscule.

For the discussion of cosmic ray effects on page 1458, lines 9-14, is it known whether this cosmic ray effect would theoretically produce a warming or cooling tendency in the most recent 50-60 years? If the effect is known to be a cooling tendency, then one does not need to argue that the magnitude is small.

For page 1458, lines 10-16, the information should be rephrased to help the reader. The aerosol effects of volcanic eruptions cannot be responsible for a 50+ -year warming trend, because that would require a trend of gradually diminishing aerosol loading from a decreasing sequence of eruptions, which has not been observed. On line 14, "eventually" should be replaced with "after a few years."

How are El Niño and climate variability related to global warming?

In the phrase "Are not caused by humans" (page 1458, line 25) it is recommended that "but their frequency and/or intensity might be affected by human greenhouse gas emissions" be added.

The statement "However, there is much uncertainty as to how climate change will affect ENSO events." on page 1459, lines 26-27, is a weak formulation that focuses on what is not known. It would be more effective to restate as, "However, these events might be affected by climate change." Similarly, page 1459, lines 30-32 provide an unnecessarily weak formulation. It would be more informative to say something to the effect that new research is shedding light on the many factors influencing how climate change affects the ENSO cycle.

The discussion about the spring transition in this FAQ does not seem particularly useful and could be omitted.

Finally, stating that "ENSO is a complex system that is controlled by hundreds of factors" (page 1459, lines 30-32) does not provide useful information to the reader interested in understanding the relationship to climate change. It would be more effective to describe studies that consider how ENSO may change under global climate change.

It is also important to note that ENSO has varied in the past, according to proxy evidence.

Is the global surface temperature record good enough to determine whether climate is changing?

This FAQ response is generally accurate, but the authors should consider that 300-year records have their own issues and are not by themselves adequate to measure global surface temperature warming. Even though this FAQ does not explicitly say that these long records have more power than they actually do, it could be inferred from the text. Instead, it might be more appropriate to emphasize records beginning in the latter part of the 19th century, from which global surface temperature estimates have been derived.

It is recommended that this FAQ answer indicate that the amount of warming varies with location, but the vast majority of Earth's surface has warmed since 1901. Since the 1950s, for the approximately 70% of Earth's surface that is ocean, all latitudes of each ocean basin have exhibited warming. The FAQ answer could also note that since 1980, the rate of warming has increased, and since 1980 each decade's global surface temperature has been successively warmer. All of the ten warmest years have occurred since 1997.

How is climate projected to change in the future?

This FAQ response emphasizes that warming will continue, but it depends on the emission scenarios, and precipitation will also increase, with wet areas getting wetter and dry areas getting drier, and heavy precipitation will be more frequent. These are all key points to convey and easily understood by the intended audience.

In the primary response to the FAQ (page 1462, lines 18-23 and page 1463, line 1) as well as in the caption of Figure A5.14, it would be clearer to add "emissions" in front of scenarios so it would not be mistaken as scenarios that are made up with no scientific basis. It is also important to point out that the U.S. temperature and precipitation changes will have large regional variation.

A robust change under warming is that more precipitation will fall as rain rather than snow, which reduces snowpack. This simple concept could be introduced in this FAQ question.

How reliable are the computer models of Earth's climate?

This FAQ is scientifically accurate, but it would benefit from further clarification of some key points.

In the text provided as the primary response to this FAQ (page 1464, lines 11-17), it may not be clear to the audience what "broad features" means. Giving an example such as jet streams, continental temperature, and precipitation patterns could help the audience to visualize what "broad features" can now be accurately reproduced by models.

On page 1464, lines 19-20, it is recommended that additional text be added: "By dividing the atmosphere, land, and ocean into smaller *spatial* units to *solve the equations*, climate models capture." Breaking up the atmosphere, land, and ocean into smaller units does not make sense unless it is noted that this is how the equations are solved.

For page 1464, lines 20-21, it is suggested that examples be added other than atmospheric-related variables, such as ocean currents and soil moisture. These are also relevant to many people and it is important to convey that climate is not all about the atmosphere.

The FAQ might also point out the many paleodata-model comparisons in which earth system models have been shown to successfully simulate major features of past climates (e.g., simulating a cold last glacial maximum, in part caused by low carbon dioxide levels at the last glacial maximum) and used to constrain estimates of climate sensitivity.

Figure A5.15 is not the right figure to use in correspondence to the sentence that cites the figure. Page 1464, lines 21-23 focuses on how well models can simulate large-scale climate features, but Figure A5.15 is about how well models can reproduce the observed trends in warming over the United States. If the main idea is to convey that climate models can now reproduce observed trends, then Figure A5.15 is not the best example. To a general audience, the disagreements between the observed and simulated temperature change in the Canadian Prairie and Mexico are enough to disqualify the statement in line 13 (page 1464) that "today's climate models can accurately reproduce broad features of past and present climate." After all, what qualifies as "accurate" is in the eyes of the beholders. Furthermore, the title of Figure A5.15, "Climate Models and Temperature Change," does not make sense. If this figure is to remain, a better title would be "Observed and model simulated temperature change." It is recommended that Figure A5.15 be replaced with a figure comparing observed and simulated large-scale features such as annual mean temperature and precipitation patterns rather than observed and simulated trends.

Can scientists project the effects of climate change for local communities?

In the sentence "(1) A statistical approach where strong, local observations are used" (page 1465, lines 21-22), it is unclear what "strong" means. Common statistical methods employ historical observations of whatever amplitude the anomalies are.

This FAQ would also benefit from some additional context. As a backdrop, it seems worthwhile to note that usually local scale conditions are strongly related to larger scale climate anomalies, whether it be calculated using statistical methods relating local to larger scale, or using a fine scale dynamical model that is driven by larger scale model guidance.

It should also be noted that regional dynamical models have biases and errors, just as do statistical methods. One can make local/regional projections, but uncertainties from global model simulations remain (climate forcing [emissions, etc.], climate sensitivity and other forms of model uncertainty, natural variability) on top of uncertainties introduced in the downscaling process.

What are key uncertainties when projecting climate change?

This FAQ is generally written at an appropriate technical level and is scientifically accurate, but some clarification is needed.

On page 1467, line 20, "climate sensitivity" should be defined for the audience or replaced by a less technical explanation such as "differences in how climate responds to doubling of greenhouse gas concentrations."

Line 13 on page 1467 could be misleading. The phrase, "different models produce slightly different projections of change," is only accurate when referring to global mean change, but audiences tend to think about climate change in terms of regional changes, so they may be misled to think that different models produce slightly different projections of regional change. This could be clarified by adding "global mean" in front of "change" as a qualifier. Also, it is suggested that "slightly different projections" be revised to "small differences in projections," since slightly is used to describe a very small difference, but certainly global mean warming between 1.5-4°C is a rather large range.

Figure A5.17 does not correspond to the description on page 1467, lines 13-14 where the figure is cited. The sentence in lines 13-14 emphasizes the range of projected outcomes due to variability, but Figure A5.17 shows the fraction of variance due to three sources of uncertainty, not just variability, and not projections but rather fraction of variance. Also, it is unclear whether Figure A5.17 is showing the fraction of variance derived for global mean temperature or some specific variable such as U.S. mean temperature. This information should be provided in the figure caption because the fraction changes with spatial scales and regions.

Is it getting warmer at the same rate everywhere? Will the warming continue?

The text provided is accurate; a few additional points would strengthen this FAQ.

It could be noted that high latitudes should warm more than other latitudes given current understanding. Also, coastal and island regions should warm less than interior continent regions.

Regarding the part of the FAQ on "Will the warming continue?," it might be useful to point out that because Earth's system still has more energy entering than leaving (oceans still warming over a very deep strata from surface to depth), global warming has not yet equilibrated to the load of increased greenhouse gases that have *already accumulated* in the atmosphere. Some greenhouse gases have long lifetimes (e.g., carbon dioxide resides in the atmosphere for a century or more). Thus, even if the *emissions* of greenhouse gases were to be sharply curtailed so as to bring them back to natural levels, Earth is committed to continued warming of more than 1°F by 2100.

Some plausible scenarios of future greenhouse gas emissions would have global surface temperature increasing by 4-9°F by 2100. Scientists think that warming in excess of 3.6°F (2°C) would have particularly dangerous consequences, placing some of these plausible scenarios into this very serious category.

What do scientists mean by the "warmest year on record"?

It is suggested that the first sentence (page 1470, lines 8-9 and 12-13) be specific and refer to the global *surface* temperature, which includes land and ocean surface.

The global surface temperature is affected by natural variability in addition to climate change; e.g., El Niño years are generally unusually warm and La Niña years are generally cool. Thus, it is unrealistic to expect each year to set a new record and this should be noted.

How can scientists project climate changes decades in the future when they cannot predict weather more than 2 weeks in advance?

Weather is an individual storm and climate is the average of many storms over many years. In the definition of climate provided, the word "average" is the key ingredient and "30 years or more" may be more appropriately explained as "over multiple years or decades." It is suggested that the text on pages 1471 be edited to read, "The climate—the average weather over multiple years to decades—varies far less."

Was there a "hiatus" in Global Warming?

Page 1473, lines 23-24, the statement is not quite right, and could by corrected by revising to, "Temporary speedups have also occurred, most notably from the early 1900s to the 1940s, and from the 1970s to late 1990s."

On page 1473, line 31 the text should distinguish an uninitialized multi-model average from initialized decadal climate predictions; the latter do show skill in simulating the slowdown. It is suggested that "less than what was expected by the models," be replaced with: "less than the average of models run with the traditional long-term increases of greenhouse gases. However, models started from specific observed conditions and run for 10 year periods successfully simulated earlier speed-ups and the more recent slowdown in the rate of surface warming."

Once again on page 1473, line 32, the different methodologies of simulating climate (uninitialized versus initialized) need to be clarified. Wording could be revised to "are more consistent with the traditional long-term model simulations with increasing human-produced greenhouse gases and have been attributed to human influence."

What is an extreme event?

The introductory sentence (page 1475, lines 6-7) is not needed given that the definition is provided in the next sentence and stated much more succinctly there.

On page 1475, line 15 the term "compounding events" is jargon and is not clear. Revised wording that could be used is, "Conversely, it is also possible for several types of extremes to occur close to the same time (e.g., a sequence of hot days that occur during dry conditions that make both worse; or several rainfall events occurring one after another that, taken together, produce flooding) that may not be considered extreme individually, but may cause."

Does global warming affect extreme weather?

The first supporting paragraph for this FAQ (page 1475, lines 26-29) would benefit from inclusion of greater context about the role of increasing greenhouse gases in the noted observation of changes in extremes. Rewording could be: "As average temperatures have warmed due to increasing human-produced greenhouse gases, extreme high temperatures have become more frequent and extreme cold temperatures less frequent. In the United States, more than twice as many daily high temperature records, as compared to low temperature records, were broken over the 2001-2012 period. With ongoing increases of greenhouse gases, the chances for extreme high temperatures will continue to increase, with the occurrence of extreme low temperatures becoming less common. However, even with much warmer average temperatures later in the century, there will still be occasional record cold snaps, though occurrences of record heat will predominate."

Including physical reasons for increasing extremes makes statements more robust. On page 1475, line 30 it is suggested that the text be modified to "Also, *because warmer air can hold more moisture,* in many areas heavy rainfall events have become."

The Knutson et al. (2017) reference cited with this FAQ is not an appropriate reference for the result it refers to.

How is climate change affecting society?

Figure A5.26 could be improved by including oceans, which would make it more consistent with the main text of the draft NCA4.

Figure A5.27 is hard to understand at first glance and thus not very effective in supporting the FAQ. The FAQ authors should consider replacing it.

It may also be worthwhile to refer readers to the "Does global warming affect extreme weather?" FAQ, particularly the discussion of extreme cold temperatures. Explaining that extreme cold weather is not proof that global warming is not happening could also be added. If an effective illustration to convey this message could be found or developed, that could also be useful.

The authors could reconsider how this FAQ is framed. The answer to "are there any benefits to climate change" is rather negative and does not frame mitigation and adaptation in the more solution-orientated nature recommended for the draft NCA4 as a whole. In some cases, mitigation and adaptation efforts may be business and innovation opportunities.

What is the social cost of carbon?

The social cost of carbon is most appropriately used to offer estimates of the economic value of changes in emissions generated by other policy interventions. It is not intended to frame mitigation development. This message should be better conveyed in the FAQ. A reference that could be used to inform this FAQ language is the "Valuing Climate Damages: Updating Estimates of the Social Cost of Carbon Dioxide" (NASEM, 2017). The social cost of carbon depends on many social values such as pure rate of time preference, relative risk aversion, etc. So, appropriate application includes a range of social cost of carbon estimates. An example of the use of the social cost of carbon would be in determining the value of increased vehicle gas mileage standards. Another type of example that could be included in the FAQ (perhaps in a more simplified form) would be siting a windmill farm that will replace fossil fuel energy. Since the effect of a single wind farm would be marginal relative to global carbon emissions, the economic value of that action can be estimated as the product of an estimate of the social cost of carbon times the reduction in carbon emissions from fossil fuel sources.

What is the difference between climate change mitigation, adaptation, and resilience?

On page 1481, lines 19-21, an example of mitigation benefits from natural systems would benefit from a longer list ("tropical forests" is not going to resonate with many U.S. residents, and thus this approach sounds like something that U.S. residents cannot participate in). A sentence pointing out that both protection and restoration of marshes, forests, and wetlands can increase carbon sequestration and storage would be more inclusive.

Is timing important when combating global warming?

It is suggested that this FAQ response lead with explaining that waiting to reduce greenhouse gas emissions will require a more rapid response later that will also be more expensive. A key point that should be added to this FAQ answer is that transient and equilibrium temperature change is driven by cumulative emissions. This could be conveyed by explaining that year-to-year emissions do not matter if the sum over a determined future is satisfied.

Are there benefits to climate change?

The answer to this FAQ should emphasize that benefits are short term and depreciate significantly in a warming world.

Are some people more vulnerable than others?

This FAQ should note that some subpopulations are more affected by environmental exposures, such as air pollution or extreme heat, in the present day. Such disparities are not limited to the future under a changing climate.

How will climate change impact economic productivity?

It is unclear what is meant by point two in the FAQ answer, "private physical capital that firms rely on to produce goods and services, such as equipment and property, will be impaired as a result of climate change" (page 1484, lines 27-28). Revision for clarity is suggested.

Can we slow or even reverse global warming?

It is suggested that the answer to this FAQ provide clearer statements first about slowing warming, and then about reversing it. Reducing the rate of emission of greenhouse gases would slow warming. That is certainly possible. To reverse warming, the amount of greenhouse gases in the atmosphere must decrease, but it is not necessary to reduce to 1750 levels, as stated on page 1485, lines 21-22. If humans stop emitting greenhouse gases, natural processes will remove carbon dioxide from the atmosphere too slowly to have any significant near-term effects on warming. Natural and technological means might be used to remove carbon dioxide (see also the FAQ related to geoengineering), but application at the necessary scale is difficult. The discussion of adaptation is off-point and should not be included in response to this FAQ.

Can geoengineering be used to remove carbon dioxide from the atmosphere or otherwise reverse global warming?

On page 1487, lines 11-16, "planting forests" is only one natural pathway for removing carbon dioxide from the atmosphere. A more general formulation would be to say that removal of carbon dioxide from the atmosphere could be undertaken by applying land management methods that increase carbon storage in forests, soils, wetlands, and other terrestrial or aquatic reservoirs.

On page 1487, lines 22-29, it is important to emphasize that solar radiation management does not reverse global warming caused by the presence of carbon dioxide and other greenhouse gases in the atmosphere (which is what the question asks). Instead, it introduces another forcing which partially cancels some of the effects of increased greenhouse gases. This is an important distinction that should be made explicitly.

Is Antarctica losing ice? What about Greenland?

This FAQ response should discuss the new ideas related to the effects of different ice sheets on different parts of the United States.

The statement: "The West Antarctic Ice Sheet, which contains enough ice to raise global sea level by 10 feet, is likely to lose ice much more quickly if its ice shelves disintegrate" (page 1491, lines 4-6), is oversimplified. In fact, much of the focus is on warming oceans eating away at the place the ice sheets go afloat in West Antarctica. A balanced statement is needed on how both warming air temperature that will make more meltwater and warming ocean could make this ice sheet collapse rapidly.

How fast are glaciers melting in Glacier National Park?

This FAQ answer is confusing and misses an important educational opportunity. Glacier recession is considered one of the important lines of evidence for climate warming and that should be emphasized. The answer should include a broader message about glaciers in general, using those in Glacier National Park as an example and less specific detail.

Some text that draws on recent findings of researchers from the U.S. Geological Survey working in Glacier National Park (GNP) as well as information from the National Snow and Ice Data Center is provided here and could be reviewed and included in the FAQ response.

- Glaciers around the world are retreating at unprecedented rates. Several ice caps, glaciers, and ice shelves have disappeared altogether this century, and many more will vanish within a matter of decades.[19] The cause is increasing temperatures and decreasing precipitation falling as snow. Glaciers retreat when melting and evaporation outpace the accumulation of new snow. In recent decades, the mountains of GNP have experienced an increase in summer temperatures and a reduction in the winter snowpack that forms and maintains glaciers. Since 1900, the mean annual temperature of GNP has increased by 1.33°C, spring and summer minimum temperatures have risen, and increases in annual precipitation have come in the form of more rain rather than snow (Pederson et al., 2010; 2011a; 2013). Mountain snowpacks now hold less water than they used to and have begun to melt at least two weeks earlier in the spring. This earlier melting alters glacier stability as well as downstream water supplies, wildlife, agriculture, and fire management.
- In a recent study, scientists looked at 39 glaciers in and around GNP and compared aerial photos and digital maps from 1966 to 2016.[20] Currently, only 26 glaciers are bigger than 25 acres, the minimum size used for defining a glacier. When GNP was established early in the last century there were an estimated 150 glaciers that were larger than 25 acres. Long term studies of glacier size have shown that the rate of melting has fluctuated in response to decade-long climate cycles and that the melting rate has risen steeply since about 1980 (Pederson et al., 2004, 2011b). Over the next 30 years, glaciologists project that most glaciers in GNP will shrink to a point where they are too small to be active glaciers, and some will disappear completely. All glaciers in the park have a severe threat of completely melting by the end of the century.

---

[19] See https://nsidc.org/cryosphere/glaciers/questions/climate.html.
[20] See https://www.usgs.gov/news/glaciers-rapidly-shrinking-and-disappearing-50-years-glacier-change-montana.

Supporting references to review when revising this FAQ response are listed in Appendix B.

How are the oceans affected by climate change?

The information in the answer to this FAQ, is accurate and in keeping with the latest available science. In general, the narrative is technical and oriented to an individual with a relatively strong science background, as it appears to make assumptions about the level of understanding the reader will have for scientific jargon (e.g., "reducing ecosystem structure and complexity" and "Atlantic Ocean's overturning circulation, known as the "Ocean Conveyor Belt"). The examples are a very good feature of the answer and additional examples could be used. Referrals back to the draft NCA4 Chapters 9, "Oceans and Marine Resources," and 24, "Northwest," are helpful. Reference to the draft NCA4 Chapter 8, "Coastal Effects," discussion of coastal ecosystems and pathways from these ecosystems to the open ocean (and to land) may also be of interest to readers and could be added.

The answer to the FAQ might be clearer to the reader if the general impacts and examples alluded to are more apparent. For instance:

- On pages 1492-1493, lines 37-38 and 1, respectively, "Dissolved CO_2 reacts with seawater and makes it more acidic. This acidification impacts marine life like shellfish and corals (see Ch. 24: Northwest)." In what ways are the examples of marine life impacted?
- On page 1493, line 3, "A warmer ocean holds less oxygen and changes the physical mixing (for example, upwelling and circulation) of oxygen in the oceans, which affects marine life." What are the examples of, or indications that, marine life has been affected?

Figure 33 could benefit from additional explanation in the caption that would link the discussion on changes in ranges of species because of climate driven ocean changes to the economic activity of fisheries (which is not really mentioned in the narrative). It should also be "A5.33" to be consistent with the numbering of other figures in the FAQ section.

What is ocean acidification and how does it affect marine life?

The pteropods example in this FAQ answer is a good one, and it especially grabs peoples' attention because they are an important food source for Pacific salmon. It would be worthwhile to add another sentence that states the link to salmon specifically. This would then provide a specific example relevant to the point in the following paragraph about ripple effects of ocean acidification in food webs.

On page 1494, lines 19-20, it would be useful to highlight corals as an especially vulnerable habitat-forming species.

How do higher carbon dioxide concentrations affect plant communities and crops?

The text is factually correct but could do more to tease out the main takeaways because as a FAQ, this text is for a busy reader that may not dig deeper. The answer could be more readable using a logical flow similar to the following:

1. Along with water, nutrients, and sunlight, carbon dioxide is one of four resources necessary for plants to grow.
2. At the level of a single plant, an increase in carbon dioxide will tend to increase or accelerate growth because of accelerated photosynthesis. Exactly how much growth stimulation will occur varies significantly from species to species.
3. However, the interaction between plants and their surrounding environment complicates the relationship. For example, [give an example, e.g., for stressed plants].
4. At the ecosystem level, the response is further complicated by the competition between species. For example, [can use the pine + poison ivy example here].
5. The expected effects of increased carbon dioxide in agricultural plants are in line with these same patterns. Crops that are not experiencing stresses from nutrients, water, or biotic stresses such as pests and disease will be expected to benefit from carbon dioxide increases. The magnitude of the effect varies greatly from crop to crop. For many crops in most U.S. regions, the benefits will likely be mostly or completely offset by increased stresses. [examples here].

This FAQ might also mention that plants often become less water stressed as the carbon dioxide concentration rises because they experience strong water-carbon tradeoffs at the leaf scale (leaf stomata let in carbon dioxide to the plant tissue [good for plant] but allow water to escape [bad for plant]). Higher atmospheric carbon dioxide concentrations allow plants to photosynthesize more with lower water losses and higher water use efficiencies.

It is suggested that the comment about nutritional quality be removed because this is still emerging science.

Comments about downstream impacts could also be a good fit here. Pollinators are mentioned, but this list could be expanded to other topics that the draft NCA4 Chapter 10, "Agriculture and Rural Communities," talks about. A topic sentence to clearly delineate for the reader that the discussion is pivoting from the strict subject of the FAQ question to closely related downstream impacts will be needed.

Is climate change affecting U.S. wildfires?

This FAQ contains some irrelevant information and omits new information that is more germane to the question and addresses recent extreme fire seasons. It is suggested that some new data/research results be reviewed to update this topic, and some simplification be made to clarify the major points.

The FAQ answer could be restructured to emphasize recent information on climate-wildfire linkages, including the attribution of recent fires to human-driven climate change. The second paragraph provides more regional detail than is necessary and should be reduced. It is also noted that much of the information is not a major topic of the draft NCA4 Chapter 6, "Forests," and the FAQ authors should consider whether better alignment is necessary.

A suggested revision to the italicized FAQ response (page 1496, lines 15-19) is: "Yes, wildfire activity occurs during periods when dry weather, adequate fuels and ignition sources co-occur. Weather determines how much area is burned, and wildfire intensity and rate of spread are closely tied to temperature, relative humidity, precipitation and wind speed. Rising temperatures and more drought have increased the frequency of wildfires as well as their size in the U.S. in recent decades (see Ch. 6: Forests, Figure A5.35)."

Long records of fire provided by tree-ring and charcoal records show that climate is the primary driver of fire on time scales ranging from years to millennia. Globally, the length of the fire season (the time of year when climate and weather conditions are conducive for fire) has increased by 19% from 1979 to 2013, and it has become significantly longer over this period in most of the United States (Jolly et al., 2015). Recent increases in the number of wildfires and area burned in most U.S. forests are a result of rising temperatures, increased drought, longer fire seasons and earlier snowmelt. Since 1985, more than 50% of the increase in area burned by wildfire in the western U.S. is attributed directly to human-caused climate change (Abatzoglou and Williams, 2016).

The frequency of large forest fires has increased since the 1970s most dramatically in the Northwest (1000%) and Northern Rocky Mountains (889%), followed by forests in the Southwest (462%), Southern Rockies (274%), and Sierra Nevada (256%) (Westerling, 2016). Dry forests in these regions account for about half of the total forest burned since 1984. The high levels of fire in these forests results from a combination of extreme weather events as well as the lasting effects of fire suppression, past logging, grazing, and invasive species in building up fuel loads. Large, high-severity fires convert unnaturally dense and structurally homogeneous dry forests to non-forest ecosystems in many places, with attendant loss of ecosystem services (Schoennagel et al., 2017).

References to review when editing this FAQ are included in Appendix B.

Appendix B. Line Comments

GENERAL

#	Page/Line	Comment
1	General	The intended audience should be more clearly explained in the Front Matter of the draft NCA4 in order to indicate the broad range of possible users of information in the report.
2	General	Similar usage of boxes across chapters would improve consistency across the report and provide readers with a clearer understanding of their general purpose.
3		Discussion of up scaling and downscaling within regional scales where data is available would be beneficial
4		The term "stressors" is used throughout the draft NCA4. It would be useful to provide a definition of what it means in the context of this report early in the draft document.
5	General	Increased usage of subject headers would help to guide readers through the chapters. This includes noting whether initial paragraphs in each chapter are intended to be a summary or an introduction. Introductions should include references while summaries do not need to include references.
6	General	It is recommended that the Front Matter (or an appendix) of the draft NCA4 provide information on how regions were determined and note more specifically the differences from the NCA3.
7	General	The Committee generally found the redundancy within chapters due to repetitious text in the executive summaries, main text, and traceable accounts to be cumbersome. Depending on how the final NCA4 report and derivative products will be structured, the NCA4 authors should consider more concise ways in which to present the material, particularly avoiding duplication between the main text and traceable accounts.
8	General	More information about the public engagement webinars and workshops for individual chapters would be beneficial to include. When was it, to whom, and how was participation solicited?

1: OVERVIEW

#	Page/Line	Comment
9	P19/L35-P20/L2	The text implies that these water impacts will occur everywhere. It seems important in these high-level messages to acknowledge that there will be regional variation in droughts and floods.
10	P25/L35	Chapter numbers are incorrect: Chapter 29 is the draft NCA4 mitigation chapter and Chapter 28 is the adaption chapter.
11	P50/L12	The authors should mention that impacts vary by region, sector, and population vulnerability.
12	P50/L20-23	This sentence could also mention carbon capture and use after "including energy efficiency."
13	P19/L22-29	Is "over the coming century" intended to mean the same as "by the end of the century"? It would seem useful to keep the same time frame for both statements.
14	P24/L10-11	The Committee recommends moving land-management practices to the end of this list of drivers, because it probably has the least impact on wildfires.
15	P24/L23	Change "resemble that" to "resemble those."
16	P26/L3-14	The upbeat tone of this paragraph sends a message of "We can do it," which is effective.
17	P29/L13-14	The graph of percent land area experiencing drought is the only short-term time series in this figure. Perhaps not surprisingly, it shows lots of variability, but the time series is not long enough to show a climate-driven trend. Can a drought-related parameter with a longer time series be used here instead? Without a longer time series or a clearer trend, this graph really is not informative to this general overview section.
18	P33/L32-33	The Alaska chapter says the climate is warming twice as fast as the global average (not MORE THAN twice as fast). Use the same time interval in the overview chapter as in the Alaska chapter and make sure the message is the same.
19	P37/L6-9	The Committee found no statement in the Alaska chapter saying that shellfish populations have been declining for 20 years and recommends the authors check this for consistency.
20	P45/L7	The time frame over which 1-4 ft sea level rise is expected is not stated. The RCP scenarios are also not stated. This makes it confusing to relate this statement to Figure 1.4.
21	P49/L33-34	The statement that emissions have stabilized is probably false. The Committee assumes that the authors mean to say that emissions are not increasing as rapidly. This should be clarified.

#	Page/Line	Comment
22	P54/L6	Delete the second occurrence of the word "access" in this sentence.

4: ENERGY SUPPLY, DELIVERY, AND DEMAND

#	Page/Line	Comment
23	P166/L23	High winds can also damage renewable energy generation and oil platforms, in addition to damaging the electricity transmission and distribution as noted here.
24	P165/L12	After the sentence about natural gas, there should be a sentence about renewable energy. After this insertion, the sentence "steps are being taken to ensure the safe and reliable" seems abrupt and unconnected to the previous sentence. Another paragraph where DOE's voluntary partnership is mentioned could be useful. Alternatively, a slight revision could work, such as an addition like, "Many actions are underway across all energy sources to ensure … but much work remains to establish an energy system that can withstand current and future climate change risks."
25	P171/L2	Maintaining additional natural gas in storage will help prevent against supply shocks and price spikes. Considering supply abundance, most modeling suggests that increased storage is not likely to increase gas or electricity prices.

5: LAND COVER AND LAND USE CHANGE

#	Page/Line	Comment
26	P189/L13	Demand for new settlements can also increase the economic cost of fire damage, as more homes move into wooded areas.
27	P189/L25	The authors could also mention policies designed to increase biofuel production.
28	P203/L16	It is not clear if the "promotes climate warming" here means that carbon dioxide emissions from deforestation will do this or if regional climates might be influenced through albedo and water and energy fluxes. Evidence presented on page 197, lines 1-10, focuses on local effects.
29	P192/L21-28	This sentence describes approaches to "increasing this carbon storage" and includes in the list of approaches "development of new generation biofuels." While the other approaches do result in increased carbon storage, there is nothing inherent in increased use of biofuels that will

#	Page/Line	Comment
		result in carbon storage by themselves, but rather would create more incentive to process and burn biomass. It is suggested that this be deleted from the list.
30	P205/L5	The confidence section addresses impacts of climate change on "urbanization in the coastal zone," but the only substantive text addressing the coastal zone is on page 195 in the draft chapter's "state of the sector" section, where links to climate change impacts are not made (only land use land cover change is described). It would be useful to make the point, probably in the "state of the sector" section, that the changes that are described for the coastal zone lead to increased impacts.
31	P196/L16	A definition for the "business-as-usual scenario" is needed, and/or reference to the description of scenarios in the draft NCA4 Front Matter.
32	P197/L14-23	There is some inconsistency in describing effects of aerosols (page 197, lines 14-15). They do not "reduce surface albedo," but rather they "increase tropospheric albedo." They reduce surface insolation. Lines 21-23 seem to get it right.
33	P199/L12-14	The expectation of conversion of irrigated agricultural land to dryland (based on the Elliott et al. 2014 reference) is reasonable in the long run (year 2100), but misses the shorter-term trend of increased irrigation and the regional differentiation that is likely in these patterns (e.g., Great Lakes versus Great Plains). See Brown, J. F., and M. S. Pervez (2014), Merging remote sensing data and national agricultural statistics to model change in irrigated agriculture, *Agricultural Systems*, 127, 28-40, doi:10.1016/j.agsy.2014.01.004.
34	P199/L16	The text says, "New policies will" The NCA4 authors do not know what new policies will do and it is not the job of the NCA4 authors to predict them. Policies certainly have affected patterns of agriculture in the past. It is recommended that this be deleted.

6: FORESTS

#	Page/Line	Comment
35	P221/L35	Insert "and spring" (revised text would say "winter and spring flooding").
36	P222/L10-15	Figure 6.5 is relevant for Pacific Northwest and some forests. Adaptation options to reduce hazardous fuels would not apply to mesic

#	Page/Line	Comment
		forests that burn infrequently and whose fires would be difficult to manage.
37	P223/L6	Add "fiber and wood products, fish and wildlife, and biodiversity" to the list of ecosystem services.
38	P223/L11-12	Provide a reference for this sentence.
39	P224/L7	Define "large-scale disturbances." Does this refer to area burned or intensity?
40	P224/L7-P225/L2	Provide temporal information for "An increase in" and "In recent years."
41	P225/L7	Define the period of "record."
42	P225/L11	Define "were lost."
43	P226/L11	Define "human welfare." Is this same as "human lives" mentioned on page 227, line 17?
44	P226/L14	Replace "would" by "will" or explain use of conditional tense.
45	P227/L13	Define "fire-prone forest ecosystems." Does this mean they have experienced frequent, low severity fires in the past or that the dominant species have adaptations to survive fires?
46	P228/L7	Specify the region or forest type where text says "some parts of the western United States."
47	P228/L14	An additional sentence is needed stating that the pattern, extent, and severity of future fires may be constrained by such breaks, provided that fire conditions do not overwhelm these barriers.
48	P228/L16-18	This sentence should clarify that prescribed burning in southern forests is an example of fire-prone managed forests.
49	P229/L12-17	More context for this sentence is needed. The fire suppression is in what type of forests and what type of insect outbreaks? Define what is meant by "reduced vigor." Is "plant host" the same as "trees"?
50	P229/L26	Insert "dry settings such as" so that revised text says "now threatens dry settings, such as the pine barrens of."
51	P230/L3	Replace "climate" with "moisture availability."
52	P230/L13	Include recent references in addition to Hicke et al. 2012. See recommended citations at the end of the line comments for this chapter.
53	P230/L13-14	This statement needs more explanation on the linkages between local short-term release of carbon dioxide and establishment of native plants.

#	Page/Line	Comment
54	P230/L19	Where do the mesic forests of the Pacific Northwest and Alaska fit into this scheme of water-limited versus energy-limited forests? Explaining this classification would be helpful.
55	P230/L22-24	Does "some locations" in this case refer to forests near urban areas? Some specificity is needed.
56	P230/L30-31	This sentence requires more information to explain the drought response of deciduous trees.
57	P230/L36	Define "vegetative" competition.
58	P231/L1	Replace "with" with "as a result of."
59	P231/L3	The text that says "these effects" is an unclear reference.
60	P231/L5	Delete the word "margins" and revise text to say "elevation ranges."
61	P231/L20-22	Increased flooding from heavy rainfall events can occur even in the absence of tree mortality. This sentence needs geographic specificity.
62	P232/L3-4	The caption needs more information. What are the colors showing? Cumulative area impacted? What in the figure shows that individual and combined disturbances are important and important in what way?
63	P232/L8-9	Explain why future conversion is expected to slow down.
64	P233/L1-2	Since one of the key issues highlighted in this chapter is potential loss of forest land, it might be useful to specify the main processes causing forest loss in western forests. Is it mainly a result of wildfire and insect disturbances or is it human conversion of forest lands to other landcovers? Is it processes similar to those projected to occur, as described in the following paragraph?
65	P233/L25	Is there more rain in the spring months as well?
66	P233/L27-29	Another result is earlier peak runoff in snow-fed headwater streams in the western U.S. "As a result…flushing of nutrients into streams has decreased." The logic for this is unclear. Do late-melting snowbanks flush more nutrients than the same quantity of runoff caused by winter rains? Is it a difference in the total amount of precipitation and runoff or does the seasonal pattern of runoff influence how much nutrient flushing occurs?
67	P233/L33	Define "climate-related changes in forest structure."
68	P233/L34-35	Explain that beetle outbreaks and wildfire create forest openings and the resulting increase in surface run-off causes higher water yield.
69	P233/L38	Define "altered."
70	P234/L5	Earlier snowmelt is primarily driven by increased temperatures, not by fires.

#	Page/Line	Comment
71	P234/L6	The word "value" (economic value?) is confusing. Perhaps replace this word with "magnitude."
72	P234/L8	Western should not be capitalized.
73	P234/L20	Insert the word "first" (text should say "can first be reduced by").
74	P234/L21	Replace "in response to the risk" with "to sustain reduced risk."
75	P234/L23-24	Explain why tree growth, carbon sequestration, and water supply are considered ecological risks. This is unclear.
76	P235/L27-29	Stand reductions to increase forest resistance/resilience to fire, insects, and drought would be effective for some but not all forest types (e.g., mesic forests, high-elevation forests, deciduous hardwood forests). This sentence needs some qualifiers.
77	P236/L3-15	The topic of this paragraph is unclear. Application of what practices and their goal should be restated. Plantation management of tree species is appropriate in forests solely managed for their wood products, but does not cover large tracts of forests on federal lands, which have multiple use mandates. The emphasis on timber extraction is not balanced.
78	P236/L20-23	Why does lower forest output lead to lower prices of products? Wouldn't supply-demand relationships lead to the opposite effect?
79	P236/L23	The cited Vaughan and Mackes (2015) study reports on a survey of Colorado forestry contractors and does not address timber output versus prices or the efficacy of adaptation treatments and incentives. The point seems counterintuitive and needs more discussion.
80	P236/L25	What is "climate-smart" forest management?
81	P237/L2-P238/L6	Where is this photo taken? Describe where beaver reintroduction is underway.
82	P238/L8	More information about the public engagement webinar would be beneficial to include. When was it, to whom, and how was participation solicited?
83	P238/L22-24	There is an unclear reference to "severe ecological disturbances" in light of reference to "other disturbances" in next sentence. More specificity or explanation of both types of disturbance is recommended.
84	P238/L24	There is an unclear reference to Hicke et al. 2016. Explain what is meant by "in combination with other disturbances."
85	P238/L25	Abatzoglou and Kolden (2013) should be cited here for western U.S. forests. Abatzoglou, J. T., and C. A. Kolden (2013), Relationships between climate and macroscale area burned in the western United States. *International Journal of Wildland Fire* 22:1003-1020.

#	Page/Line	Comment
86	P238/L25	Give time frame for phrase "in recent years."
87	P238/L32	"Re-burns" have not been mentioned previously, so they should be defined and explained.
88	P238/L32-33	The potential for subsequent fires also depends on fire-fighting and postfire management actions. Discussion of this is recommended.
89	P239/L3	This is the first mention of "historical range of variability." It should be defined, referenced and discussed.
90	P239/L17-20	Define "gradual climate change" in this paragraph and note that the examples cited come from North America and Europe. This statement should link to Chapter 7, "Ecosystems, Ecosystem Services, and Biodiversity."
91	P239/L23-25	This is an unclear statement. More specificity where the text says "can affect suitable habitat" is needed. Define or replace "elevation range margins" with "elevational ranges" or define "elevation range margins."
92	P239/L37-38	The cited Caldwell et al. 2016 study is a local study in North Carolina, which attributes lower runoff to a combination of changes to climate, structure, and species composition, not just climate. Reference to Roman et al. 2015 seems inappropriate for this point.
93	P240/L16-18	Describe the long-term observations (and location) that are referred to here. It is a cryptic point.
94	P240/L33	This citation should say "McCarthy et al. 2006." Wear and Coulson 2015 seems to be a better citation for this point.
95	P240/L36	Specify the locations ("some locations") where this may be true.
96	P241/L1-4	This sentence somewhat overstates the conclusion of Kurz et al. 2008, which looks at the loss of carbon from mountain pine beetle outbreaks in British Columbia. It is a single study.
97	P241/L7	It would be helpful to provide some context for this statement as it does not apply equally to all forests (e.g., Pacific Coastal forests, eastern forests).
98	P241/L35-36	Identify the location of the studies in the Cascades and state the duration of the impact (decreased reflectivity, etc.).
99	P241/L37	Explain how the conclusions of Luce et al. 2012 were informed by the more recent studies of Gleason et al. 2016 and Cooper et al. 2016. Luce et al. is an older study in a different region.
100	P242/L6-9	There are surprisingly few long-term (tree-ring) studies on carbon dioxide effects on tree growth in older trees. The authors may consider citing Gedalof and Berg (2010).

#	Page/Line	Comment
		Gedalof, Z., and A. A. Berg (2010), Tree ring evidence for limited direct CO2 fertilization of forests over the 20th century, Global Biogeochemical Cycles, 24(3), doi:10.1029/2009GB003699.
101	P242/L6	Define "altered disturbance patterns." Note that some of the uncertainty comes from the resolution of downscaled climate model projections.
102	P242/L8	This is unclear. What particular future trends in natural and socioeconomic systems are critical?
103	P242/L20	The statement of "reduced tree growth and carbon storage observations" needs specific context.
104	P242/L30-35	This planning effort should be identified by name, location, and the organization doing the planning. The list is too general to be helpful without more information.
105	P242/L36-P243/L6	This discussion would benefit from literature on fire resilience efforts (i.e., living with fire). The cited Schoennagel et al. 2017 is a good start, but see additional suggestions in the reference list following the line comments for this chapter.
106	P243/L14	The reference to Worrall et al. 2013 seems inappropriate, since it addresses aspen decline.
107	P243/L17-18	There is an unclear reference: "more abundant [than what]"?
108	P243/L24	Add "conservation of biodiversity or endangered species" and "protection of plants/places of special importance to indigenous peoples" to the list of specific actions.

Chapter 6-Suggested References

FIRE TRENDS AND LEGACY EFFECTS	• Barnett, K., S. A. Parks, C. Miller, and H. T. Naughton (2016), Beyond fuel treatment effectiveness: Characterizing Interactions between fire and treatments in the US, *Forests, 7*(237), 1-12. • Dennison, P. E., S. C. Brewer, J. D. Arnold, and M. A. Moritz (2014), Large wildfire trends in the western United States, 1984-2011, *Geophysical Research Letters, 41*(8), 2928-2933. • Fulé, P. Z., J. E. Crouse, J. P. Roccaforte, and E. L. Kalies (2012), Do thinning and/or burning treatments in western USA ponderosa or Jeffrey pine-dominated forests help restore natural fire behavior? *Forest Ecology and Management, 269*, 68-81. • Hudak, A. T., I. Rickert, P. Morgan, E. Strand, S. A. Lewis, P. R. Robichaud, C. Hoffman, and Z. A. Holden (2011), Review of fuel treatment effectiveness in forests and rangelands and a case study from the 2007 megafires in central Idaho USA, General Technical

	Report RMRS-GTR-252. U.S. Department of Agriculture, Forest Service, Rocky Mountain Research Station, Fort Collins, CO. • Kalies, E. L., and L. L. Y. Kent (2016), Tamm Review: Are fuel treatments effective at achieving ecological and social objectives? *Forest Ecology and Management, 375*, 84-95. • Kemp, K. B., P. E. Higuera, and P. Morgan (2015), Fire legacies impact conifer regeneration across environmental gradients in the U.S. northern Rockies, *Landscape Ecology, 41*(3), 619-636. • Picotte, J. J., B. Peterson, G. Meier, and S. M. Howard (2016), 1984-2010 trends in fire burn severity and area for the conterminous US, *International Journal of Wildland Fire, 25*(4), 413-420.
RAPID FOREST CHANGE AND DISTURBANCE SYNERGIES	• Hansen, W. D., F. S. Chapin, H. T. Naughton, T. S. Rupp, and D. Verbyla (2016), Forest-landscape structure mediates effects of a spruce bark beetle (Dendroctonus rufipennis) outbreak on subsequent likelihood of burning in Alaskan boreal forest, *Forest Ecology and Management, 369*, 38-46. • Harvey, B. J., D. C. Donato, and M. G. Turner (2016), Drivers and trends in landscape patterns of stand-replacing fire in forests of the US Northern Rocky Mountains (1984-2010), *Landscape Ecology, 31*(10), 2367-2383. • Hart, S. J., T. Schoennagel, T. T. Veblen, and T. B. Chapman (2015), Area burned in the western United States is unaffected by recent mountain pine beetle outbreaks, *Proceedings of the National Academy of Sciences,* doi:10.1073/pnas.1424037112.
LONG-TERM FOREST CHANGE	• Dobrowski, S. Z., J. Abatzoglou, A. K. Swanson, J. A. Greenberg, A. R. Mynsberge, Z. A. Holden, and M. K. Schwartz (2013), The climate velocity of the contiguous United States during the 20th century, *Global Change Biology*, *19,* 241-251. doi:10.1111/gcb.12026. • Hudiburg, T. W., P. E. Higuera, and J. A. Hicke (2017), Fire-regime variability impacts forest carbon dynamics for centuries to millennia, *Biogeosciences, 14,* 3873-3882, https://doi.org/10.5194/bg-14-3873-2017. • Iverson, L. R., A. M. Prasad, S. N. Matthews, and M. Peters (2008), Estimating potential habitat for 134 eastern US tree species under six climate scenarios, *Forest Ecology and Management, 254,* 390-406. • Rehfeldt, G. E., N. L. Crookston, M. V. Warwell, and J. S. Evans (2006), Empirical analyses of plant-climate relationships for the western United States, *International Journal of Plant Sciences, 167,* 1123-1150.
FIRE ADAPTATION AND RISK ASSESSMENT	• Calkin, D. E., J. D. Cohen, M. A. Finney, and M. P. Thompson (2014), How risk management can prevent future wildfire disasters in the wildland-urban interface, *Proceedings of the National Academy of Sciences, 111*(2), 746-751.

- Calkin, D. E., M. P. Thompson, and M. A. Finney (2015), Negative consequences of positive feedbacks in US wildfire management, *Forest Ecosystems, 2*(1), 1-10.
- Chang, T., A. J. Hansen, and N. Piekielek (2014), Patterns and variability of projected bioclimatic habitat for Pinus albicaulis in the Greater Yellowstone area, *PLoS One*, doi:10.1371/journal.pone.0111669.
- Chapin, F. S., S. F. Trainor, O. Huntington, A. L. Lovecraft, E. Zavaleta, D. C. Natcher, A. D. McGuire, J. L. Nelson, L. Ray, M. Calef, N. Fresco, H. Huntington, T. S. Rupp, L. Dewilde, and R. L. Naylor (2008), Increasing wildfire in Alaska's boreal forest: Pathways to potential solutions of a wicked problem, *Bioscience, 58*(6), 531. (Alaskan forests are not considered in chapter)
- Flatley, W. T., and P. Z. Fulé (2016), Are historical fire regimes compatible with future climate? Implications for forest restoration, *Ecosphere, 7*(10), e01471.
- Harvey, B. J., D. C. Donato, and M. G. Turner (2016), High and dry: Post-fire tree seedling establishment in subalpine forests decreases with post-fire drought and large stand-replacing burn patches, *Global Ecology and Biogeography, 25*(6), 655-669.
- Headwaters Economics (2016), Land Use Planning to Reduce Wildfire Risk: Lessons from Five Western Cities. Available at: headwaterseconomics.org/wphw/wpcontent/uploads/Planning_Lessons_Full_Report_Web.pdf
- Hessburg, P. F., D. J. Churchill, A. J. Larson, R. D. Haugo, C. Miller, T. A. Spies, M. P. North, N. A. Povak, R. T. Belote, P. H. Singleton, W. L. Gaines, R. E. Keane, G. H. Aplet, S. L. Stephens, P. Morgan, P. A. Bisson, B. E. Rieman, R. B. Salter, and G. H. Reeves (2015), Restoring fire-prone Inland Pacific landscapes: Seven core principles, *Landscape Ecology, 30*(10), 1805-1835.
- Hessburg, P. F., T. A. Spies, D. A. Perry, C. N. Skinner, A. H. Taylor, P. M. Brown, S. L. Stephens, A. J. Larson, D. J. Churchill, N. A. Povak, P. H. Singleton, B. McComb, W. J. Zielinski, B. M. Collins, R. B. Salter, J. J. Keane, J. F. Franklin, and G. Riegel (2016), Tamm Review: Management of mixed-severity fire regime forests in Oregon, Washington, and Northern California, *Forest Ecology and Management, 366,* 221-250.
- Hobbs, R. J., E. Higgs, C. M. Hall, P. Bridgewater, F. S. Chapin, E. C. Ellis, J. J. Ewel, L. M. Hallett, J. Harris, K. B. Hulvey, S. T. Jackson, P. L. Kennedy, C. Kueffer, L. Lach, T. C. Lantz, A. E. Lugo, J. Mascaro, S. D. Murphy, C. R. Nelson, M. P. Perring, D. M. Richardson, T. R. Seastedt, R. J. Standish, B. M. Starzomski, K. N. Suding, P. M. Tognetti, L. Yakob, and L. Yung (2014), Managing the whole landscape: Historical, hybrid, and novel ecosystems, *Frontiers in Ecology and Environment, 12*(10), 557-564.

	• NRAP (2017), *Northern Rockies Adaptation Partnership: Vulnerability Assessment Summaries.* Available at http://adaptationpartners.org/nrap/docs/NRAP_vulnerability_assessment.pdf, accessed April 25, 2018. • Iverson, L. R., A. M. Prasad, S. N. Matthews, and M. Peters (2008), Estimating potential habitat for 134 eastern US tree species under six climate scenarios, *Forest Ecology and Management, 254,* 390-406. • Moritz, M. A., E. Batllori, R. A. Bradstock, A. M. Gill, J. Handmer, P. F. Hessburg, J. Leonard, S. McCaffrey, D. C. Odion, T. Schoennagel, and A. D. Syphard (2014), Learning to coexist with wildfire, *Nature*, *515*(7525), 58-66. • Rasker, R. (2015), Resolving the increasing risk from wildfires in the American West, *Solutions, 6*(2), 55-62. • Schoennagel, T., J. K. Balch, H. Brenkert-Smith, P. E. Dennison, B. J. Harvey, M. A. Krawchuk, N. Mietkiewicz, P. Morgan, M. A. Moritz, R. Rasker, M. G. Turner, and C. Whitlock (2017), Adapt to more wildfire in western North American forests as climate changes, *Proceedings of the National Academy of Sciences*, doi:10.1073/pnas.1617464114. • Thompson, M. P., P. Bowden, A. Brough, J. H. Scott, J. Gilbertson-Day, A. Taylor, J. Anderson, and J. R. Haas (2016), Application of wildfire risk assessment results to wildfire response planning in the southern Sierra Nevada, California, USA, *Forests, 7*(3), 64-86.
PALEOECOLOGY	• Cook, E. R., R. Seager, M. A. Cane, and D. W. Stahle (2007), North American drought: reconstructions, causes, and consequences, *Earth-Science Reviews, 81,* 93-134. • Gavin, D. G., and L.B. Brubaker (2014), *Late Pleistocene and Holocene environmental change on the Olympic Peninsula, Washington*, Springer Verlag, Dordrecht. • Heyerdahl, E. K., P. Morgan, and J. P. Riser (2008), Multi-season climate synchronized historical fires in dry forests (1650-1900), Northern Rockies, USA. *Ecology*, *89*(3), 705-716. • Higuera, P. E., C. E. Briles, and C. Whitlock (2014), Fire-regime complacency and sensitivity to centennial- through millennial-scale climate change in Rocky Mountain subalpine forests, Colorado, U.S.A., *Journal of Ecology, 102,* 1429-1441. • Jackson, S. T., J. L. Betancourt, R. K. Booth, and S. T. Gray (2009), Ecology and the ratchet of events: climate variability, niche dimensions, and species distributions, *Proceedings of the National Academy of Sciences, 106*(2), 19,685-19,692. • Marlon, J. R., P. J. Bartlein, D. G. Gavin, C. J. Long, R. S. Anderson, C. E. Briles, K. J. Brown, D. Colombaroli, D. J. Hallett, M. J. Power, E. A. Scharf, and M. K. Walsh (2012), Long-term perspective on wildfires in the Western USA, *Proceedings of the National Academy of Sciences, 109,* E535-E543.

- Mclauchlan, K. K., P. E. Higuera, D. G. Gavin, S. S. Perakis, M. C. Mack, H. Alexander, J. Battles, F. Biondi, B. Buma, D. Colombaroli, S. K. Enders, D. R. Engstrom, F. S. Hu, J. R. Marlon, J. Marshall, M. Mcglone, J. L. Morris, L. E. Nave, B. Shuman, E. A. H. Smithwick, D. H. Urrego, D. A. Wardle, C. J. Williams, and J. J. Williams (2014), Reconstructing disturbances and their biogeochemical consequences over multiple timescales, *Bioscience, 64,* 105-116.
- Odion, D. C., C. T. Hanson, A. Arsenault, W. L. Baker, D. A. DellaSala, R. L. Hutto, W. Klenner, M. A. Moritz, R. L. Sherriff, T. T. Veblen, and M. A. Williams (2014), Examining historical and current mixed-conifer forests of western North America, *PLoS ONE, 9*(2), e87852, doi:10.1371/journal.pone.0087852.
- Pederson, N., J. M. Dyer, R. W. McEwan, A. E. Hessl, C. J. Mock, D. A. Orwig, H. E. Rieder, and B. I. Cook (2014), The legacy of episodic climatic events in shaping temperate, broadleaf forests. *Ecological Monographs, 84*(4), 599-620 doi:10.1890/13-1025.1.
- Shuman, B., A. K. Henderson, C. Plank, I. Stefanova, and S. S. Ziegler (2009), Woodland-to-forest transition during prolonged drought in Minnesota after ca. AD 1300, *Ecology, 90*(10), 2792-2807.
- Swetnam, T. W., and J. L. Betancourt (1998), Mesoscale disturbance and ecological response to decadal climatic variability in the American Southwest, *Journal of Climate, 11,* 3128-3147.
- Whitlock, C., P. E. Higuera, D. B. McWethy, and C. E. Briles (2010), Paleoecological perspectives on fire ecology: revisiting the fire-regime concept, *The Open Ecology Journal, 3,* 6-23.
- Whitlock, C. (1992), Vegetational and climatic history of the Pacific Northwest during the last 20,000 years: implications for understanding present-day biodiversity, *The Northwest Environmental Journal, 8,* 5-28, 1992.

7: ECOSYSTEMS, ECOSYSTEM SERVICES, AND BIODIVERSITY

#	Page/Line	Comment
109	P257/L14	Insert "-" (revised text would say "large marine-ecosystem scales").
110	P257/L21-25	Not all of the topics listed here are discussed in the chapters in the context of the many people, communities, and economies that depend on the services.
111	P259/L4	Delete "_"
112	P259/L5	The clause ", which include" has an unclear reference. If this is a definition of ecosystems, it should be clearly called out.
113	P259/L6	Ecosystem services should be clearly defined.

#	Page/Line	Comment
114	P259/L9	The United States is sometimes abbreviated as U.S. and elsewhere it is spelled out. This should be discussed consistently across the chapter (and report).
115	P259/L14	Insert "in the future" after "change" so that the text reads, "change in the future still."
116	P259/L21-31	By focusing on the "state" of biodiversity and ecosystems, the NCA4 authors place a singular focus on impacts on species, communities, etc. There is virtually no mention of what ecosystem services are likely to be affected, as flows from "stock" (biodiversity and natural ecosystem components) to people, which is the definition of ecosystem services.
117	P259/L24	The word "phenology" should be moved to come after "migration."
118	P259/L29-39	"This" is an unclear reference.
119	P259/L30-31	Insert "or past response" so that the text reads, "modeling its individual components or past response."
120	P259/L34	Provide a specific example of a shift in phenology and population performance.
121	P259/L35-P260/L1	The cited Cleland et al. 2012, Willis et al. 2010, Chuine 2010, Zimova et al. 2017, are not in the reference list and are inconsistently formatted with other references in the chapter.
122	P260/L6	The word "or" should be replaced by "and/or."
123	P260/L7-9	Insert a time span of observation for these statements about range changes and provide some specific examples.
124	P260/L10-12	Range shifts are different among terrestrial species as well, so it is not clear what in this statement is unique to marine plankton.
125	P260/L15-16	This assertion that species will move north and up is too simplistic. Several studies in the western U.S., for example, show complex range changes that best track changes in effective moisture.
126	P260/L7-8	Over what time have communities shifted ranges? This observation is not helpful without more information about the community, the time span of observation, or the location.
127	P260/L18	This is unclear: "species' responses" to what? Is this a reference to their range changes?
128	P260/L20	Define "other stressors" or remove this phrase if it is covered in the list already.
129	P260/L21	It is unclear what is meant by "topography and the interaction of different climate aspects."

#	Page/Line	Comment
130	P260/L28-29	The text needs a time span for "increasing rate of introduction of non-native species globally."
131	P260/L30	This is unclear: "costs" of what?
132	P260/L32	Define "novel communities."
133	P260/L32-33	The references provided in support of this point are inappropriate. A specific example is also needed.
134	P261/L2	Do the authors mean to say "native species" instead of "existing non-native species"? This should be clarified.
135	P261/L10	Suggest replacing "manifested through" with "as evidenced by."
136	P261/L11-12	The list is not parallel. It is suggested that "the ecosystem services they support" be replaced with "ecosystem services."
137	P261/L12	Instead of stating, "Nationally" it is suggested that "Across the U.S." be used.
138	P261/L12	This is unclear: "starting earlier" than when?
139	P261/L15-16	It is unclear what the temporal baseline for this observation is. This should be explained.
140	P261/L21-23	Provide an example to support this sentence.
141	P261/L28	Replace "predators" with "consumers."
142	P261/L35	Replace "are able" with "will" in both places.
143	P262/L5-8	Rewrite "stressors increase stress." Explain how a human-caused stressor "decreases the overall gene pool." Provide an example.
144	P262/L6	It is unclear what is meant by "natural systems."
145	P262/L13-16	Provide a specific example for some of these groups.
146	P262/L16-19	Provide an example of an evolutionary change in response to climate change.
147	P262/L20-24	These two sentences seem contradictory. Evolution will not counteract the negative effects of climate change and evolution will have negative effects. This point needs clarification (and an example).
148	P262/L34	Delete "(."
149	P262/L37-38	It is unclear what "other factors" are. Please specify.
150	P263/L5	Replace ", which has" with ", and this change."
151	P263/L5	Hyphenate "mid-latitude."

#	Page/Line	Comment
152	P263/L6-7	This sentence about mixed evidence is cryptic without more information. Also note the time span for decreased productivity (last century).
153	P263/L12	Define the timeframe for "recent observations."
154	P263/L13-15	For clarity, explain the link between increased productivity and changes in fisheries catch.
155	P263/L20-24	This statement needs an example. Higher energetic needs would be a direct result of warmer temperatures, drought, and extreme events. This should be factored into this sentence, which focuses only on biotic interactions. Also define "resource mismatches."
156	P263/L24-27	This statement would benefit from an example.
157	P263/L38	This last sentence in this paragraph is cryptic. What are the debates? It would be good to cite Barnosky et al. 2017 here. Barnosky, A. D., et al. (2017), Merging paleobiology with conservation biology to guide the future of terrestrial ecosystems, *Science*, *355*(6325), doi:10.1126/science.aah4787.
158	P264/L5-37	The template is not followed for the Regional Roll-Up and the second paragraph is a mish mash of unrelated topics. Some of the statements are questionable (e.g., attributing salmonid declines to climate change versus bears). It is suggested that reference be made to information provided in particular regional chapters. For instance, Tolan and Fisher, 2009 is cited in the draft NCA4 Chapter 23.
159	P264/L26	Endangered fisheries are the result of land use change as much as climate change.
160	P265/L22	Give an example to support this statement about shifts in phenology.
161	P265/L29-37	This paragraph refers to climate change impacts on ecosystem services, but the information is too general to be evaluated. Some specificity and examples would help.
162	P266/L6-12	The treatment of U.S. federal agency policies/actions is vague. Provide specific example(s) to make it real for the reader.
163	P266/L8	Insert "," so that revised text would say "food conditions, they."
164	P266/L23-38	References on climate resilience should be included here.
165	P266/L31	Fix the citation: "Service 2013."
166	P266/L32	The authors should review and consider citing the guidance document Stein et al., 2014.

#	Page/Line	Comment
		Stein, B.A., P. Glick, N. Edelson, and A. Staudt (eds.). 2014. Climate-Smart Conservation: Putting Adaptation Principles into Practice. National Wildlife Federation, Washington, DC.
167	P266/L36	Define "holistic ecosystem-based approaches."
168	P267/L4-5	It is unclear what is meant by "mitigate the harmful impacts of current and future resource management challenges."
169	P267/L5	The word "agencies" should be in lower case.
170	P268/L1-2	Cite references for the statement regarding range shift consequences.
171	P268/L28	Insert time span to support "species respond to climate change."
172	P268/L32	Insert time span for "experienced range shifts."
173	P268/L32-35	Define climate velocity. Also cite Dobrowski and Parks, 2016, for discussion of climate change velocity/exposure in mountainous areas. Dobrowski, S. Z., and S. A. Parks (2016), Climate change velocity underestimates climate change exposure in mountainous regions, *Nature Communications*, *7*, 12349, doi:10.1038/ncomms12349. https://www.nature.com/articles/ncomms12349#supplementary-information.
174	P268/L34	Explain why text says "can be greater," rather than "is."
175	P269/L1-2	The first sentence needs some specific case studies for this to make sense. Provide some direct links to climate change and references to specific documents.
176	P269/L3	Explain why timber production will shift as a result of climate change.
177	P269/L5-6	Define "tragedy of the commons," which will likely be unfamiliar to a general audience.
178	P269/L7-9	This statement would benefit from an example and reference to regional chapters that discuss Indigenous issues.
179	P269/L8	Delete "both."
180	P269/L8-10	More citations should be provided to support this sentence. There are more recent papers than Graves, 2008 and ones that cover full breadth of statement.
181	P269/L15	Provide an example of "climate-induced phenological change" to support this sentence.
182	P269/L20-24	This interesting observation about migratory birds needs a time span for the data.

#	Page/Line	Comment
183	P269/L24-25	It is unclear what is meant by "sufficiently advance migratory phenology."
184	P269/L30	Provide an example of altered pollinator-prey relationships.
185	P269/L35	It is unclear what a "climatological expectation" is and what the time frame for this observation is.
186	P270/L4	Define "standing genetic diversity."
187	P270/L5	This is unclear: "more gradual" than what?
188	P270/L6-9	This observation suffers by the lack of specific examples. How is adaptation to climate change identified?
189	P270/L10	Replace "other non climate-related stressors" with "non-climatic stressors" and define this term.
190	P270/L11	Replace "predictions" with "projections." Specify the critical climate variables (see draft NCA4 Chapter 2).
191	P270/L20	Define "communities." In this case, does this refer to human communities?
192	P270/L21	Replace "communities" with "economies."
193	P270/L22	Explain: "economically vulnerable to what," and how does this limit their response?
194	P270/L23	Provide an example of an invasive species that is having an economic impact in the face of climate change.
195	P270/L25-27	The two economic statistics are nonparallel, so it is difficult to compare them.
196	P270/L28-37	This section would benefit from a case study under the key message.
197	P270/L28	Insert "-" (revised text would say "land-use change").
198	P270/L29	Insert "ecological" so that revised text states "ecological communities."
199	P270/L32	It is unclear what is meant by "behavioral mechanisms."
200	P270/L33	It is unclear what is meant by "specific traits."
201	P271/L3	Clarify: "major outbreaks" of what?
202	P271/L5-7	It would be useful to provide an example of how novel species are making the noted changes.
203	P271/L9	This sentence seems out of place.
204	P271/L13-23	These two sentences make no sense and do not follow each other.
205	P271/L24	Explain "advanced modeling techniques."

#	Page/Line	Comment
206	P271/L28	Monitoring should be included as a shortcoming and critical need.
207	P271/L29	It is unclear what "under" means.
208	P271/L36	Insert space between "variation" and "(Jeong."
209	P272/L3	It is unclear why there is a long list of citations here.
210	P272/L5	What is the demonstrated "uniquity of local adaptation" that is referred to? This is important and more detail should be provided.
211	P272/L8	What is the year of the publication?
212	P272/L8	Explain the two case studies.
213	P272/L10	For the phrase, "involved changes in the timing of migration," what time period is being discussed?
214	P272/L13	The use of the term "novel" is used differently here than the rest of the text-here it refers to invaded areas. The term should be used consistently, or clearly defined when used differently.
215	P272/L13	Provide an example of how available methods have been used to estimate risk.
216	P272/L14	What are the emerging technologies noted here?
217	P272/L17	Define "novel sectors and livelihoods."
218	P272/L21	Provide an example of how novel ecosystem transitions may result. The paleoecological literature might be helpful in this regard.
219	P272/L28	It is unclear how the references are related to the sentence.
220	P272/L36	This text repeats line 20-24 but with different references.
221	P273/L18	There are no case studies to demonstrate how well natural resource management and adaptation strategies are working now or need to be refined in order to work better.
222	P273/L27-28	Define "seed sourcing" and "assisted migration" for a general audience.
223	P273/L31-32	No case studies are provided in the text to support this statement.
224	P273/L32-35	The statement would be strengthened with some examples.
225	P274/L3	Vermont grassland systems are human-created landscapes. What is their value for biodiversity or conservation? Describe the likely phenological shifts related to climate change.
226	P274/L5	Replace "has" with "have."

#	Page/Line	Comment
227	P274/L15-17	This reference to resistance/resilience is not well discussed in the key message section and should be clarified. How is resistance/resilience with respect to climate change defined and what are the compromises posed by invasive species?
228	P274/L20-23	This long list of citation needs some examples in order for the reader to understand the state of uncertainty and the potential for large changes in this regard.
229	P274/L25	What is the reference to "recent" here? Some of the citations go back to before the NCA3, so they are not new in relation to the development of this draft fourth assessment.
230	P274/L35	Plans for state and public-private partnerships should be discussed.
231	P275/L14	It is unclear why is the word "could" is used here.

9: OCEANS AND MARINE RESOURCES

#	Page/Line	Comment
232	P332/L19-20	A reference and some elaboration is needed on the statement that ocean acidification or low oxygen events can lead to technological adaptation.
233	P334/L5	The population and percentage value listed differ slightly from that provided in the draft NCA4 Chapter 8. The chapters should report the information using the same values for consistency.
234	P334/L25-37	Why are no new references on ocean acidification since the NCA3 cited? The discussion is basic information, rather than new data on the status or trends.
235	P335/L7-10	A reference is needed for the "dead zone" in the Gulf of Mexico.
236	P336/L4	Burrows et al. 2014 would also be an appropriate reference to include here. Burrows, M. T., et al. (2014), Geographical limits to species-range shifts are suggested by climate velocity, *Nature*, *507*(7493), 492-495, doi:10.1038/nature12976.

10: AGRICULTURE AND RURAL COMMUNITIES

#	Page/Line	Comment
237	P373/L27-P375/L5	Adaptation through land-use change is acknowledged as an option (page 375, line 5), but not addressed with any specific examples. Consider

#	Page/Line	Comment
		including one or more of the following instances where more detail or examples could be helpful: (1) page 375, where the need for adaptation in the long run should be more strongly stated (Pugh et al., 2016), (2) page 373, line 27 and page 374, line 1, the amount of cropland in 2012 was down substantially from the 1950s (Brown et al., 2005) in response to change markets, technology, and policy. Less productive areas that have been abandoned could be brought back into production as an adaptation measure, though new ownership and land-use/livelihood patterns may reduce the capacity for reversion to agriculture as patterns of productivity change. Loss of cropland to urbanization, for example, limits reversion to cropland. (3) Paragraph beginning on page 375, line 16, paragraph addresses adaptation by a variety of means, but not by land-use change. Retirement of agricultural land or conversion to pasture at the productivity margins is a form of adaptation that has been happening over centuries. As noted above, it may be the key form of adaptation necessary in the long run. This will have significant effects on rural communities (both those where cropping declines and those where it increases). Evidence from econometric studies could be included, such as Burke and Emerick (2016), Feng et al. (2015), and Burke and Emerick (2016) (cited in draft chapter). (4) Address bioenergy and bioenergy with carbon capture as mitigation options and the additional interconnected stresses a massive scale up in bioenergy for mitigation would likely cause. See new citations at the end of the line comments for this chapter.
238	P373/L12-16	Remove "Food service, eating and drinking places," "Food and beverage stores," and "Textile, apparel, and leather manufacturing" from Figure 10.1. Their magnitude and distant relevance to the chapter (i.e., these are largely urban enterprises) distract from the message.
239	P374/L6-11	It is true and well-documented that agriculture has become more efficient over the last few decades. However, consider following up the statement about reducing agricultural and environmental footprint with a comment about some of the remaining major environmental footprint issues to be addressed (e.g., eutrophication in Great Lakes and the Gulf of Mexico).
240	P375/L16-P376/L9	The comment about the effectiveness of existing adaptation strategies given continued productivity growth does not consider the possibility that growth could have been much faster with even better adaptation. Given the concurrent trends of continuous equipment/genetic improvements, which were not designed as climate mitigation strategies, it is probably not safe to assume from the last couple of decades of increasing productivity that the sector is particularly skilled at climate change adaptation already.

#	Page/Line	Comment
241	P375L13-15	Challenges to food security should acknowledge changing diets along with population growth as drivers of this challenge.
242	P376/L6-9	Research needs are referenced here, but few other places in the chapter. It would be highly valuable to comment more systematically on views of what the key research challenges are (related to this chapter's scope) in the next few years, as appropriate for inclusion in this draft report.
243	P376/L35-P377/L5	Mention of increasing irrigation as possible adaptation should reference observed increases, while also acknowledging the importance of water resource limitations in the future, as the text does. Data from Brown et al. 2014 could give a sense of scale. Also, acknowledge the regional variability of resource limitations (some basins are in much better or worse shape than others).
244	P377/L27-P378/L7	Consider adding a comment on yield quality in addition to quantity, e.g., nutritional quality of crops under climate change scenarios. This may qualify as an emerging topic to watch. The draft NCA4 Chapter 23 cites Myers et al. 2017 on this.
245	P377/L29-31	"The demand for higher crop productivity under climate change has contributed to advancements in crop genetics in recent years." Do the provided references support this? Robust crop breeding programs certainly are a mechanism for continually adjusting crop genetics to recent weather conditions and thereby ought to help agriculture progressively adjust to some types of climate change. In some cases, breeding programs have more directly targeted traits that help with drought resilience, etc., as noted in the text. But that is different than climate change adaptation being a direct motivator of recent crop genetic advancements.
246	P377/L1-5	This statement is probably true without climate change also-recharge is not keeping up with withdrawals in a lot of places. Climate change might accelerate this, but the chapter needs to recognize the baseline.
247	P378/L20-P379/L6	There may be somewhat of an overemphasis of the chapter text on irrigation as measured by the amount of text in the chapter devoted to this relative to U.S. acreage percentage using irrigation. Maybe it is an appropriate ratio based on economic importance? If so, the authors could state the outsized economic importance of irrigated acres. Either way, it is not clear that it is appropriate to have the sole case study box in this section to be about groundwater-fed irrigation of High Plains row crops.
248	P378/L20-P379/L6	"[T]he Ogallala aquifer is a nonrenewable resource." This requires more qualification, as there is a major difference in conditions from north to south across the aquifer. Groundwater recharge rates in the northern portion are quite high and approximately capable of sustaining current

#	Page/Line	Comment
		irrigation rates (although river flows have suffered). As recent supporting evidence, groundwater levels in Nebraska recovered well following the record drawdowns during the 2012 drought year. In contrast, the central and southern portions have low recharge and should mostly be considered non-renewable resources.
249	P378/L8-9	Define "climate-smart agriculture" and reference the origin of this term.
250	P381/L33-38	Clarification is needed in the statement about migration of the feeding industry. Is the expectation that industry in the southern Great Plains and Atlantic coastal plain will contract towards the northern Great Plains and upper midwest? Is there a reference describing the current trend?
251	P381/L25	"Similar arguments" is too vague of a reference. To what arguments does it refer?
252	P382/L21-39	The nutrient loss pathway discussion focuses on erosion, but leaching (with or without tile drains) is another major loss pathway. The word leaching does not appear in the chapter. Relatedly, the combination of warmer springtime soils and increased rainfall intensity would seem to have the potential to increase nitrogen leaching under U.S. row crops, potentially reversing a trend of increasing nitrogen use efficiency over time in the sector. The cited Rosenzweig et al. 2014 paper mentions nitrogen losses as potentially exacerbating yield reductions. Pesticide leaching is also a risk. Consider adding comments on these mechanisms. A related point is that agronomic management practice shifts that are designed to mitigate climate impacts could also unintentionally exacerbate downstream impacts, for example, adding more fertilizers to insure against the yield losses described by the Rosenzweig et al. 2014 citation included in the draft NCA4, could potentially add to waterway discharges if they are not coupled with precision application technologies or similar.
253	P383/L5	"[S]uch declarations" are not defined. A reference is provided in the traceable accounts (see the reference to "billion dollar natural disasters" on page 390, line 1), but it is unclear if they are referring to the same thing. More explanation is needed.
254	P383/L4-8	This paragraph seems out of scope since it is not focusing on agricultural and rural communities.
255	P384/L23	This is confusing wording. Change to "are less likely to exist and more loosely enforced."
256	P385/L38	"TPF" should be "TFP."

#	Page/Line	Comment
257	P386/L8-14	Is the moisture/fire relationship for the period 1600-1800 safely applicable to the 2000s given all of the land use and land cover differences?
258	P386/L8-14	Evidence for the fire prevalence impact on agriculture and how economically impactful it is should be added.
259	P386/L8-14	The cited Margolis et al. 2017 is locally focused on northern New Mexico. Additional references to support the statement more broadly would be beneficial.
260	P386/L16	Dai and Zhao 2017 may have found positive trends in drought indices (not negative, as stated). It is suggested that the authors revisit this literature and confirm/correct.
261	P387/L27-P388/L14	Consider separating and making explicit the main types of research evidence about climate change impacts on crops. For example, (1) observational field studies, (2) experimental field studies, and (3) modeling studies. Also consider citing by name an example from each, such as the cited Hatfield et al. 2017 and several Lobell studies for (1), free air carbon dioxide experiment (FACE) studies for (2), and the Agricultural Model Intercomparison and Improvement Project (AGMIP) for (3).
262	P390/L7-13	This section justifies comments on mitigation capabilities but not on the impacts themselves.
263	P402/L7-10	This reference is used four times in the chapter and is a link to a non-technical overview. Reference to the full work should be included.

Suggested References

- Brown, Daniel G., et al. "Rural land-use trends in the conterminous United States, 1950–2000." *Ecological Applications* 15.6 (2005): 1851-1863.
- Brown, J. F., & Pervez, M. S. (2014). Merging remote sensing data and national agricultural statistics to model change in irrigated agriculture. *Agricultural Systems* 127, 28-40.
- Burke, Marshall, and Kyle Emerick. "Adaptation to climate change: Evidence from US agriculture." *American Economic Journal: Economic Policy* 8.3 (2016): 106-140
- Feng, Shuaizhang, Michael Oppenheimer, and Wolfram Schlenker. 2015. "Weather Anomalies, Crop Yields, and Migration in the US Corn Belt." http://www.columbia.edu/~ws2162/articles/FengOppenheimerSchlenker.pdf.

- Jin, Zhenong, et al. "Increasing drought and diminishing benefits of elevated carbon dioxide for soybean yields across the US Midwest." *Global change biology* (2017).
- Lobell, David B., et al. "Greater sensitivity to drought accompanies maize yield increase in the US Midwest." *Science* 344.6183 (2014): 516-519.
- Myers, S.S., et al., Climate Change and Global Food Systems: Potential Impacts on Food Security and Undernutrition. *Annu Rev Public Health*, 2017. 38: p. 259-277.
- Porter, J. R., Xie, L., Challinor, A. J., Cochrane, K., Howden, S. M., Iqbal, M. M., ... Travasso, M. I. (2014). Food security and food production systems. In Climate Change 2014: Impacts, Adaptation, and Vulnerability (pp. 485-533). Cambridge University Press. http://curis.ku.dk/ws/files/131829514/Chapter_7._Food_security....pdf.
- Pugh, T. A. M., et al. "Climate analogues suggest limited potential for intensification of production on current croplands under climate change." *Nature Communications* 7 (2016): 12608.
- Urban, Daniel W., Justin Sheffield, and David B. Lobell. "The impacts of future climate and carbon dioxide changes on the average and variability of US maize yields under two emission scenarios." *Environmental Research Letters* 10.4 (2015): 045003.
- Wang, Zhaozhi, et al. "Modeling the impacts of climate change on nitrogen losses and crop yield in a subsurface drained field." *Climatic Change* 129.1-2 (2015): 323-335.

12: TRANSPORTATION

#	Page/Line	Comment
264	P451/L1	It would be more appropriate to say that transportation is "A" not "THE" backbone, along with communications and energy.
265	P451/L19	The transportation system is "INTERdependent with" other systems, not just dependent on.
266	P452/L5	The notion of "a new transportation paradigm" is raised, but is not defined. Do these new approaches (transit-oriented development (TOD), autonomous vehicles, and shared mobility) actually constitute a new paradigm?
267	P452/L34	The NCA4 authors should check on whether the state of the science indicates that there is too much uncertainty in forecasts about the levels of the Great Lakes to say that lower levels WILL limit boat traffic.

#	Page/Line	Comment
268	P456/L1-3	This is a theoretical statement about heat impacts. This is a well-known process to be sure, but it would be strengthened with empirical observations of the effect.
269	P460/L9	Examples of resiliency measures that have or could be taken should be provided. This is not obvious from the text.
270	P468/L28	"[C]omprise" should be "compromise."
271	P468/L37	It would be useful to relate the physical effect of buckling at >90 degrees to the number of projects days under the scenarios.
272	P470/L11-15	Empirical and modeling evidence seems to be slim for this "high confidence." Presumably, high confidence could be based on engineering studies, but it would be stronger with observations and models.
273	P470/L32	This statement about low-income people being less likely to evacuate comes from another assessment report. The primary evidence should be provided.
274	P471/L1	The meaning of "broad constituencies in suburban areas" is vague and requires clarification.
275	P471/L9-10	Why give examples of Colorado and Iowa? In what way are they representative?

15: TRIBAL AND INDIGENOUS COMMUNITIES

#	Page/Line	Comment
276	P548/L26-33	The introduction in general, and this paragraph in particular, would benefit from references to support these statements. If this initial section is intended as an overview summary rather than an introduction, a heading should be added to indicate this. It is an opportunity to inform the general readership of the NCA4 about publications that document the important points raised in the introduction. For example, literature that documents increased rates of mood and anxiety disorders is needed. This statement is repeated in the "results" (draft NCA4 page 555, lines 32-34). Do the NCA4 authors intend this to be a finding of the assessment (in which case, page 555 and the associated key message is a good place to present it) or as background (in which case, the introduction would be the better place)? More general statements have been well documented in the literature, which should be cited in the introduction.
277	P548/L34-37	See previous comment #274. Additional references that would allow readers to further explore the content of this paragraph are needed.

#	Page/Line	Comment
278	P552/L3-7	Key Message 1 is important. However, with a superficial reading, it would seem to apply to non-tribal as well as tribal entities. The logic would be more compelling if the chapter explained why these vulnerabilities are often greater for tribal than for non-tribal entities. Text that references Figure 15.2 might be an appropriate place to make these explanations. Box 15.1 does a good job of providing these types of explanations with respect to social determinants of indigenous health.
279	P555/L6	Replace "that can that" with "that can."
280	P555/L38-P556/L3	This section needs citations and more thorough explanation.
281	P556/L4-11	This paragraph does a good job of providing citations for each key statement.
282	P558/L2-7	This is an excellent example of the types of statistics that are valuable to present in the report.
283	P559/L5-7	The NCA4 authors should be a bit more explicit about these issues. It seems like an important point, but it is so general that it does not provide a lot of insight. Is this because the governance issues are so heterogeneous from place to place that one cannot generalize, i.e., that greater future assessment will be necessary to make progress, or is it that the authors chose not to provide the details?
284	P547-563	General comment: The chapter is clearly written and addresses the important climate change impacts on indigenous peoples in a rigorous fashion. In some instances, the same points are made in the introduction and results, making it unclear which aspects are part of this assessment and which are general background. The inclination is to suggest general background papers be cited in the introduction and that the results focus on evaluation of papers that bring new information to this assessment. See also comment #274.
285	P550/L7	The term "federally recognized Tribes" appears without definition or characterization of important distinctions regarding legal/policy standing and political relationships with other groups of indigenous peoples, non-federally recognized tribes, state recognized tribes, and other groups of self-recognized peoples. It is recommended that the term be defined.
286	P550/L9	This characterization of the trust responsibility is misleading. The trust responsibility has two major components: (1) fiduciary obligations of the U.S. as trustee for the management of the Indian estate, the funds and resources entrusted to its care; and (2) the duty to support tribal self-determination and role in the American system of governance. What is described as "trust responsibility" is in fact a policy recognition that federally recognized tribes are political sovereigns that are to be

#	Page/Line	Comment
		treated in accordance with protocols appropriate for government-to-government relations. There are several statutes and policies that describe responsibilities for interacting with tribal governments, including consultation on matters that affect their rights and interests.
287	P550/L15-22	Consider relocating this paragraph to the start of the Executive Summary.
288	P551/L13-14	Figure 15.1 does not reflect "models of adaptation," but rather indicates locations of tribal involvement in climate change initiatives.
289	P551/Figure 15.1	Adaptation is not well covered in the web links provided in this figure. The first web link simply refers back to this draft NCA4 chapter; the second and third web links describe climate impacts but say relatively little about ongoing adaptation activities.
290	P551/L16	The word "cumulatively" should be "collectively."
291	P551/L19-20	It is recommended that the last sentence be deleted because it serves little substantive purpose. Although islands are depicted in Figure 15.1, the chapter text does not substantively discuss many of the unique issues faced by indigenous peoples in these locations.
292	P552/L15	Change "comprise" to "contain."
293	P552/L15	The validity of this statement is questioned here. No source for this statistic is presented. Most tribal hatcheries do not produce fish that are not listed under the ESA.
294	P553/L7	Authors should consider citing Parsons et al. 2017. Parsons, M., C. Brown, J. Nalau, and K. Fisher (2017), Assessing adaptive capacity and adaptation: insights from Samoan tourism operators, *Climate and Development*, 1-20, doi:10.1080/17565529.2017.1410082.
295	P553/L14	There are not just regulatory responses and impacts on arts and crafts income. Failure to enforce laws and regulations (e.g., clean air and water, energy efficiency), conflicting missions among fragmented agencies, and failure to allow for traditional uses and management practices, etc. all affect availability (access, abundance, and productivity) of many resources for food security, subsistence, medicines, and commercial and ceremonial use.
296	P553/L23-30	The pervasive role of federal agencies and persistent remnants of paternalistic policies are embodied in the manuals, rules, and regulations. These policies and procedures that are relied upon to administer fiduciary trust responsibilities are major barriers to adaption and development of tribal resources/economies. See Energy Act, NIFRMA, Indian Ag Act, etc. Additionally, note persistent poverty,

#	Page/Line	Comment
		lack of infrastructure, and isolation (e.g., 14% lack access to electricity [Energy Information Administration], and only a small percentage have access to broadband Internet).
297	P555/L5	This discussion would benefit from references to environmental/climate justice to address disproportionate distribution of impacts to the economically disadvantaged and populations of color.
298	P555/L19	The word "undermine" should be changed to "alter." These relationships are not limited to humans and animals, but to all aspects of the environment, plants, water, soils, air, etc.
299	P555/L23-31	An aspect of climate change which is important, but not directly covered, involves scarcity of resources and competition, not just among indigenous peoples' communities, but also as a result of recreational and commercial use of these resources by a growing population. Additionally, trespassing on tribal lands, environmental degradation, and some reserved rights to areas off reservation are also affected.
300	P556/L11	Impacts also accrue to traditions and practices, not just to sacred sites.
301	P556/L27	The discussion of adaptation is overly broad. Consideration should be given to the utility of distinguishing between different types of adaptation experienced by indigenous peoples. Certainly the ability to adapt to changing natural environments over millenia of pre-European contact, when indigenous communities were much more mobile within ancestral territories, differs markedly from the trauma of populations devastated by disease and forced adaptation to changes in natural and political environments resulting from the imposition of Western political, legal, and economic systems of property ownership, dispossession, relocation, and assimilation policies. These challenges differ markedly from those being faced today by indigenous peoples as they strive to contend with the necessity to deal with the pace and intensity of adaptation, preparation, and mitigation measures needed to respond to climate change. The ability of indigenous peoples to anticipate and respond to climate change is affected by economic, political, and legal considerations that severely constrain their abilities to consider and undertake alternative actions.
302	P556/L36	Revise text to say "multi-generational accumulation and transfer of knowledge."
303	P557/L25	It is recommended that "within recognized areas where reserved hunting, fishing, gathering, and trapping rights can be exercised" be added.
304	P557/L25	A major impact that is not mentioned is reduced abundance and productivity due to environmental degradation and development that affect ecological processes.

#	Page/Line	Comment
305	P557/L38	Should section 1110 of the Sandy Recovery Improvement Act or Stafford Act be referenced?
306	P558/L2	"President" should be capitalized.
307	P558/L20-25 & P559/L17-28	The relevance of forced relocation due to climate change should be clarified. Rather than "frameworks"—whatever those might be—three larger problems are apparent: (1) the difficulty of maintaining community/cultural continuity of place and environment for communities under relocation; (2) the lack of resources to support physical relocation, including aspects of governance—taxation, regulation, etc.; and (3) the impacts on the communities and environments receiving relocated communities.

21: MIDWEST

#	Page/Line	Comment
308	P844/L8-15	The chapter should address (at least major categories of) adaptation in the agricultural sector.
309	P850/L1	The impacts of transitions from an extreme drought year to an extreme flood year are mentioned, but no information is provided on whether this is expected in the future. Will there be more, fewer, the same, or is it unknown? If it is unknown, it seems unnecessary to mention the impact.
310	P851/L6-7	It would be useful to comment on increasing irrigation use in some parts of the region.
311	P857/L1515	Climate factors interact with one another and they interact with land use land cover patterns. This should be noted.
312	P863/L29-P864/L1	The implication is that decreasing lake levels can be expected. The evidence is not clear on this. Of course, if they do decrease, there will be increasing shipping costs. It would be more appropriate to say something about the science of whether or not we can expect them to decrease. Reduced ice cover is more clear.
313	P887/L19-20	Provide references for the following claim: "It is clear, however, that flood frequency on major rivers in the Midwest has increased in recent decades."

22: NORTHERN GREAT PLAINS

#	Page/Line	Comment
314	P916/L15	"[G]eographical migration of agricultural practices" seems like jargon. Clarification on what exactly this means is needed.
315	P917/L12	The Northern Great Plains extends to Wyoming and Montana and significantly wetter conditions are not forecasted for the western parts of these states.
316	P919/L1-7	It would be helpful to show other basic climate projections for the region (seasonal/annual temperature, seasonal/annual precipitation projections).
317	P921/L2	These three geographic features should be shown on a map and there should be some identification beyond these features. The three features are also not clearly identified in the text (Red River Valley, Upper Missouri River Basin, and the third being the mountains of Wyoming and Montana?).
318	P921/L11	Define "alpine water dynamics." Is this precipitation in headwater systems?
319	P921/L18	This is a very sparsely populated region.
320	P921/L18-23	The NCA4 authors should mention dryland wheat production in Montana. The reference to arid to semiarid climate of this region requires some climate information.
321	P921/L21	Delete "."
322	P921/L22	It is unclear what is meant by the "western portion of the region," but it is managed for agriculture, forestry, grazing, and recreation.
323	P921/L35	Insert "-" such that revised text says "long-term."
324	P922/L12	The statement that it is among the most arid in the Nation should be supported with some precipitation data and an identification of what area is referred to specifically.
325	P922/L24	Delete "a" so that text reads, "representing new and unprecedented."
326	P922/L25	Is this a reference specifically to Glacier National Park?
327	P922/L35	Add Montana to the list.
328	P924/L11	In Figure 22.2, an explanation is needed for why Snow Water Equivalent for Average March is used instead of the usual April 1 Snow Water Equivalent.
329	P925/L5	Explain "high degree of variability." Does this mean annual, seasonal, or spatial? Not only is the variability high, but the uncertainty is also high.

#	Page/Line	Comment
330	P925/L8-9	Delete "in the future."
331	P925/L15	These projections also apply to the northwestern portion (i.e., northern Montana).
332	P925/L17	Replace "which" with "and projected changes."
333	P925/L19	In addition to agriculture and energy production, the list should be expanded to include human health, streamflow and temperatures, snow melt, fires, etc.
334	P925/L24	Nebraska has more humid-continental climate than where? Does this statement apply year-round or to seasonal climate conditions?
335	P925/L25-26	Does this statement about reservoir and groundwater storage apply everywhere in the region or only in the eastern part?
336	P927/L7	It is unclear what is meant by "essential vegetation heterogeneity."
337	P927/L19	A more detailed discussion of producer decision-making would be helpful, since climate change is only one factor. Refer to Whitlock et al. 2017 provided in the references at the end of this chapter review and the draft NCA4 Chapter 5 for a start.
338	P927/L28	Provide information about summer precipitation.
339	P927/L28	This paragraph should also cite the increase in extreme precipitation events. For example, a study of hail: Brimelow et al. 2017. Brimelow, J. C., W. R. Burrows, and J. M. Hanesiak (2017), The changing hail threat over North America in response to anthropogenic climate change, *Nature Climate Change*, *7*(7), 516-522, doi:10.1038/nclimate3321.
340	P927/L30	Insert "(."
341	P927/L34-P928/L17	These projections do not apply to the entire Northern Great Plains region, which extends to western Montana. The authors need to be geographically specific. Item 1 should provide a season. Item 6 does not apply to the western region, where livestock will experience greater stress as a result of late-season drought and high temperatures.
342	P928/L29	Consider changing "would" to "will."
343	P928/L30	Define "sustainability" in this case, or replace with "resilience."
344	P929/L19	Explain the Collaborative Adaptive Rangeland Management experiment and how it relates to The Nature Conservancy's Matador Ranch.
345	P929/L25	The phrase: "under each which were used" is awkward and should be reworded.
346	P930/L8	To clarify, $4.9 billion was spent in the Northern Great Plains states?

#	Page/Line	Comment
347	P930/L28	More regional citations that should be reviewed/cited include: Al-Chokhachy, R., D. Schmetterling, C. Clancy, P. Saffel, R. Kovach, L. Nyce, B. Liermann, W. Fredenberg, and R. Pierce (2016), Are brown trout replacing or displacing bull trout populations in a changing climate?, *Canadian Journal of Fisheries and Aquatic Sciences*, *73*(9), 1395-1404, doi:10.1139/cjfas-2015-0293. Giersch, J. J., S. Hotaling, R. P. Kovach, L. A. Jones, and C. C. Muhlfeld (2017), Climate-induced glacier and snow loss imperils alpine stream insects, *Global Change Biology*, *23*(7), 2577-2589, doi:10.1111/gcb.13565. Muhlfeld, C. C., R. P. Kovach, L. A. Jones, R. Al-Chokhachy, M. C. Boyer, R. F. Leary, W. H. Lowe, G. Luikart, and F. W. Allendorf (2014), Invasive hybridization in a threatened species is accelerated by climate change, *Nature Climate Change*, *4*(7), 620-624, doi:10.1038/nclimate2252.
348	P930/L29	This line needs a citation to support the content.
349	P931/L10	Insert "the" so the text reads, "last day of the snow." Do not capitalize spring.
350	P932/L5-9	Other initiatives should be mentioned such as the Greater Yellowstone Coordinating Committee (representing all federal agencies in the Greater Yellowstone Area) that has several climate-change related initiatives, and Crown of the Continent Partnership, which is similarly looking at climate change impacts on headwater streams.
351	P932/L22	This should say "northern and eastern Montana."
352	P933/L77	Table 22.3 should include information from Montana, which represents about 30% of the prairie pothole region.
353	P936/L6	Stating "things" is not very specific. It is suggested that this be changed to "initiatives" or "programs."
354	P936/L8-19	The authors should mention that the largest emitters of greenhouse gases in this region are coal-fired power plants and that Wyoming and North Dakota are the highest emitters of greenhouse gas per person in the nation.
355	P937/L8-12	A description of Department of Energy (DOE) supported efforts for carbon capture and sequestration should be described.
356	P938/L7	Why use "some" if they are "among." Delete "some."
357	P938/L13	It is unclear what the difference is between climate and seasonality changes. Is this meant to say changes in annual and seasonal climate?
358	P938/L18-21	This sentence is poorly written. Rewrite for clarity and parallelism. Suggested edit: "changes in hydrology, phenology, availability of

#	Page/Line	Comment
		traditional plant-based foods, bear migration and hibernation cycles, as well as the health of [whitebark?] pine? There is also a mismatch between traditional stories of past climate and current climate conditions."
359	P938/L22-25	These are general statements (e.g., no salmon in this region), where more specific examples would be helpful.
360	P938/L24	Delete ",".
361	P938/L31	State which language is quoted.
362	P938/L33	There is a misspelling of the word "Because."
363	P939/L3	Replace "were" with "was."
364	P939/L9	Provide more specificity for the statement "changes to temperature and water cycles."
365	P939/L10	It is suggested that "increasing livestock stress" be added to the list.
366	P940/L6	Provide more specificity for "projected to damage infrastructure." What types of climate change, for what infrastructure, and where is this happening?
367	P940/L22	Define "colonial/postcolonial."
368	P940/L30	There is a misspelling of the word "Dakota."
369	P942/L9	The term "South-central" should be in lower case, as in "south-central" or "south central."
370	P942/L15	Delete ",".
371	P942/L24	There should be no capitalization of the word "chokecherry."
372	P942/L30	There should be no capitalization of the word "olive."
373	P943/L3-5	For Figure 22.6's caption, there should be no capitalization of the word "olive."
374	P943/L3-5	This inset is very hard to evaluate. What is the current distribution of Russian olives and what do the colors and the scale mean? The warm colors may suggest that there is a projected decrease in Russian olives to many readers.
375	P945/L10	Explain why there are no authors from Montana, South Dakota, North Dakota or Wyoming. This does not seem like appropriate representation for a regional assessment and this will reduce its creditability among stakeholders.
376	P945/L36	Replace "montane west" with "western mountains."
377	P946/L10	In addition to year-to-year variability, there is increased seasonal variability. See the draft NCA4 Chapter 2 and the Montana Climate

#	Page/Line	Comment
		Assessment (Whitlock et al. 2017 in the reference list at the end of this chapter's line comments).
378	P946/L29	Specificity is needed as not all parts of the region will show an increase in productivity.
379	P946/L30	What weeds in particular are competitive? Greater specificity is needed.
380	P946/L36	It would be helpful for a call-out of these particular studies, especially those studies specific to the Northern Great Plains.
381	P947/L15	The reference to "geographical migration of agricultural practices and enterprises" is part of Key Message 2, but it is not discussed.
382	P947/L36-37	This sentence is awkward and needs greater clarity.
383	P948/L5-6	Agricultural land-use change is not a function of climate change. A more nuanced discussion of this point and the factors that shape land-use decisions is needed.
384	P948/L14	What is more important for skiing is that climate change will shorten the ski season, which has economic consequences for the skiing industry.
385	P948/L18	This list should mention the impacts affecting cold-water fisheries, e.g., more invasive species, warmer water temperatures, and lower flow.
386	P948/L24	In addition to disease, upriver movement of warm-water fish and displacement of cold-water species should be mentioned.
387	P949/L10	There are not only climate-induced changes to agricultural land-use, but also changes to the wetlands themselves through late-season drying, early snowmelt, etc.
388	P949/L16	Replace "is" with "are."
389	P950/L10	Specific examples, or at least references, are needed here.
390	P950/L22	This is the first reference to the Columbia River Basin, which represents a very small part of the Northern Great Plains region.
391	P950/L27-28	This sentence about biofuel production is very cryptic. What other biofuels are used? Specify some of the climate change impacts.
392	P951/L32	It would be helpful to list state-level climate assessments for additional information.

Chapter 22-Suggested References

- Bathke, D.J., R. J. Oglesby, C. M. Rowe, and D. A. Wilhite (2014), *Understanding and Assessing Climate Change: Implications for Nebraska,* University of Nebraska-Lincoln.

- Bekkerman, A., G. W. Brester, and M. Taylor (2016), Forecasting a moving target: the roles of quality and timing for determining northern US wheat basis, *Journal of Agricultural and Resource Economics, 41,* 35-41.
- Brimelow, J. C., W. R. Burrows, and J. M. Hanesiak (2017), The changing hail threat over North America in response to anthropogenic climate change, *Nature Climate Change, 7,* 516-22.
- Lanning, S. P., K. Kephart, G. R. Carlson, J. E. Eckhoff, R. N. Stougaard, D. M. Wichman, J. M. Martin, and L. E. Talbert (2010), Climatic change and agronomic performance of hard red spring wheat from 1950 to 2007, *Crop Science, 50,* 835-841.
- Pederson, G. T., S. T. Gray, C. A. Woodhouse, J. L. Betancourt, D. B. Fagre, J. S. Littell, E. Watson, B. H. Luckman, and L. J. Graumlich (2012), The unusual nature of recent snowpack declines in the North American cordillera, *Science,* 333(6040), 332-335.
- Plowright, R. I., P. C. Cross, G. M. Tabor, E. Almberg, L. Bienen, and P. J. Hudson (2012), Climate change and infectious disease dynamics, in *New Directions in Conservation Medicine: Applied Cases of Ecological Health,* A. Alonso Aguirre, Richard Ostfeld, and Peter Daszak, eds., Oxford University Press, Oxford, UK.
- Whitlock, C., W. Cross, B. Maxwell, N. Silverman, and A. A. Wade (2017), Montana Climate Assessment, Montana State University and University of Montana, Montana Institute on Ecosystems, Bozeman and Missoula, MT, doi:10.15788/m2ww8w.
- Dobrowski, S. Z., J. Abatzoglou, A. K. Swanson, J. A. Greenberg, A. R. Mynsberge, Z. A. Holden, and M. K. Schwartz (2013), The climate velocity of the contiguous United States during the 20th century, *Global Change Biology, 19,* 241-251. doi:10.1111/gcb.12026.

23: SOUTHERN GREAT PLAINS

#	Page/Line	Comment
393	P966/L30	Brimelow et al. 2017 should be cited for hail threat. Brimelow, J. C., W. R. Burrows, and J. M. Hanesiak (2017), The changing hail threat over North America in response to anthropogenic climate change, *Nature Climate Change*, *7*(7), 516-522, doi:10.1038/nclimate3321.
394	P970/L33-P971/L4	Add references to statements in the last two paragraphs of the box regarding changes to projected frequencies and intensities.
395	P971/L7	The time span for this observation needs to be stated.
396	P973/L22-38	The paragraph on the drought in 2011-2015 and its resulting economic impacts is an interesting one, but the text is not currently clear about

#	Page/Line	Comment
		what the intended takeaways are: (a) an indicator of the kinds of things we expect to increase in frequency in the future with climate change, (b) an episode that is attributable to climate change, or (c) other? Please specify. Also, consider this as a potentially suitable case study box.
397	P973/L28	Replace "coal plant" with "coal-fired power plant."
398	P974/L2-P975/L4	The main point of this box is not clear. Is this intended to just point out an example where climate early warning information exists? It would be a more powerful example if a successful utilization of this information by a stakeholder could be described.
399	P977/L18	Insert "that" so the text reads, "role that climate."
400	P978/L34	The Oklahoma Mesonet is referenced here. In general, the three states comprising this region have very good state/local monitoring systems, which, in some ways, have been a model for other regions. Consider mentioning climate monitoring investments in the region in the context of either trend identification or adaptive capacity-building.
401	P979/L6	Replace "Nation" with "the U.S."
402	P980/L21-32	The Box 23.4 discussion on the El Paso desalination plant is quite limited and its takeaways are not clear. How much are costs reduced relative to seawater desalination given the brackish waters? Also, consider revising the box title since the desalination plant is being discussed as one mechanism in a broader suite of methodologies for achieving water security.
403	P982/L12-15	Whooping cranes vulnerabilities being used as an example of species range changes should be referenced in the draft NCA4 Ecosystems chapter (Chapter 7) if discussed here.
404	P983/L25	Where is the "Texas bay" located? Is this same as the Texas Gulf Coast?
405	P983/L31	The words "ground water" should be one word: "groundwater."
406	P988/L10	It is not clear how Phytophthora is influenced by an increase in invasive species. Is this a non-native fungus?
407	P991/L29-30	It would be helpful to include more information on the types of individuals that attend the engagement workshop and how they engaged in the process.
408	P991/L38	The difference between "doctors, academicians, researchers and scientists" is not clear. Please clarify. Are they medical doctors? Are the researchers from federal and state agencies?
409	P994/L7-8	The description of confidence and likelihood for Key Message 3 makes the point that habitat created by invasive species due to climate changes has improved populations of other species. This may be in reference to

#	Page/Line	Comment
		increased invasive species following flooding and the increase in detrimental fungal species, but it is not entirely clear. This is mentioned only on page 988, lines 8-12, but the point is not clearly developed and seemed like a minor observation.
410	P994/L22-26	The description of evidence base for Key Message 4 makes reference to Chapter 7 on the point of increased microbial and chemical contamination of crops and water in agricultural environments. That message is not currently clear in Chapter 7.
411	P994/L7-8	Insert "including fungi" at the end of the sentence.

24: NORTHWEST

#	Page/Line	Comment
412	P1017/L23-33	Also cite consequences of temperature/range shifts for fisheries (not all are negative—some fish species are/will enter fishing areas where they did not previously occur). See references in the draft NCA4 Chapter 9 and elsewhere.
413	P1018/L10	Change "effect" to "affect."
414	P1018/L1-7	Include documented literature on range shift effects on fisheries/management (not just species), e.g., Ianelli et al. 2016 and other references in the draft NCA4 Chapter 9.
415	P1018/L18-21	Provide a citation for the good point on variation in adaptive capacity throughout the region.
416	P1025/L20-21	Cite Ianelli et al. 2016 and Seung and Ianelli 2016 in this chapter, which are referenced in the draft NCA4 Chapter 9).
417	P1026/L28-29	Cite mitigation and other co-benefits from climate-based species/wildlife management as another emerging area.
418	P1031/L11-12	The text should say "has reduced the impact of sea level rise for *some areas* in the Northwest."
419	P1031/L21-P1032/L2	Include hybrid "green and gray" infrastructure approaches as part of the emerging issues.

25: SOUTHWEST

#	Page/Line	Comment
420	P1088/L8-10	Mention why the significant technology sector in the region cares about climate change (or how climate change affects the sector).

#	Page/Line	Comment
421	P1088/L17-18	Water supply varies not only with precipitation, but also with withdrawals/use, as mentioned in the preceding sentence.
422	P1092/L19-20	Regarding Oroville Dam spill risks, reduced water supply and maintenance required for aging infrastructure are two risks mentioned in the text. Another major spill risk is the risk to human lives/property/infrastructure from catastrophic failure.
423	P1092/L38	Text should say "rather than being forced *to* use it immediately."
424	P1094/L2-8	The middle sentences on drought should be rephrased to focus on fire impacts. Start with "forests have dried," driving wildfire increase, and then include an explanation of how droughts have contributed to forest drying.
425	P1094/L21	Text should say "carbon, *in* California ecosystems."
426	P1098/L36-37	Text should say "in natural*ly* acidic upwellings."
427	P1099/L3-4	The text needs a citation for the economic risks of ocean acidification to the shellfish industry.
428	P1100/L8	The legend should be consistent with the true color of the temperature line in Figure 25.4. It looks brown in the draft NCA4, but the caption states that it is black.
429	P1102/L2-13	The paragraph starts with a discussion of how tribes are adapting; the paragraph ends with an example of how vulnerable they are to decreasing water supply. This text needs edits for internal consistency. Perhaps the two points could be merged with other paragraph(s).
430	P1103/L14-15	Briefly state the cause of the increased cost in shifting from hydroelectric to fossil fuels in California (see the cited Gleick, 2015 reference).
431	P1103/L36-39	This sentence needs editing for length and clarity.
432	P1104/L8	Text should say "Energy *production* causes the."
433	P1104/L11-14	This sentence needs editing for length and clarity.
434	P1104/L14-15	What economic damages would carbon emission reductions cut and how? This needs a brief explanation.
435	P1114/L21	A citation is needed for the portion of this sentence referring to the increase in flooding. This is the first mention of flooding related to this key message and is strangely stated as part of the documentation in the "drought" key message. More detail on this point can be found in the main text review of this draft NCA4 chapter.

#	Page/Line	Comment
436	P1117/L12	An incomplete sentence ends with "to diminishing."

Chapter 25 Suggested References

SUPPORT FOR KEY MESSAGE 5	• Johnstone, J. A., and T. E. Dawson (2010), Climatic context and ecological implications of summer fog decline in the coast redwood region, *Proceedings of the National Academy of Sciences* 107(10), 4533-4538, doi:10.1073/pnas.0915062107. • Iacobellis, S.F., and D.R. Cayan, 2013: The variability of California summertime marine stratus: Impacts on surface air temperatures. *Journal of Geophysical Research—Atmospheres* 118, 9105-9122, doi:10.1002/jgrd.50652. • Schwartz, R.E., A. Gershunov, S.F. Iacobellis, and D.R. Cayan, 2014: North American west coast summer low cloudiness: Broadscale variability associated with sea surface temperature. *Geophysical Research Letters* 41(9), 3307-3314, doi:10.1002/2014GL059825. • Torregrosa, A., C. Combs, and J. Peters (2016), GOES-derived fog and low cloud indices for coastal north and central California ecological analyses, *Earth and Space Science* 3, 46-67, doi:10.1002/2015EA000119.
SUPPORT FOR KEY MESSAGE 7	• Guirguis, K, Gershunov A, Tardy A, Basu R. 2014. The impact of recent heat waves on human health in California. *Journal of Applied Meteorology and Climatology* 53:3-19. doi:10.1175/jamc-d-13-0130.1. • Gershunov, A, Guirguis K. 2012. California heat waves in the present and future. *Geophysical Research Letters* 39 doi:10.1029/2012gl052979. • Guirguis, K., A. Gershunov, D.R. Cayan and D. Pierce, 2017: Heat wave probability in the changing climate of the Southwest US. *Climate Dynamics* doi:10.1007/s00382-017-3850-3.
WEB-ACCESSIBLE ITEMS FOR URBAN CLIMATE ISSUES AND BARRIERS TO ADAPTATION	• The Climate Change in the Los Angeles Region Project; http://research.atmos.ucla.edu/csrl/LA_project_summary.html • San Diego, 2050 is Calling. How Will We Answer?; http://www.adaptationclearinghouse.org/resources/san-diego-2050-is-calling-how-will-we-answer.html • Memo: Key Countywide Survey Findings on San Diego County Residents' Knowledge of and Attitudes Toward Climate Change; http://catcher.sandiego.edu/items/climate/2017%20SDCEP%20External%20Summary%20Memo%20D2.pdf • Climate Education and Opportunities: Key Informant Interviews for 2017 Collective Impact Summit;

	http://catcher.sandiego.edu/items/climate/20170830_Key%20Informant%20Interviews%20Report_final.pdf • San Francisco Climate Action Strategy; https://sfenvironment.org/cas
EXAMPLES OF RESOURCES FOR ADAPTATION AND PLANNING	• California Climate Change Assessment; http://resources.ca.gov/climate/safeguarding/research • Assessment of Climate Change in the Southwest United States; http://www.swcarr.arizona.edu • Climate Change in Colorado; http://wwa.colorado.edu/climate/co2014report • San Diego Regional Climate Collaboration; https://www.sdclimatecollaborative.org • Los Angeles Regional Collaborative; http://www.laregionalcollaborative.com • Capitol Region Climate Readiness Collaborative; http://climatereadiness.info • Bay Area Climate Collaborative; https://en.wikipedia.org/wiki/Bay_Area_Climate_Collaborative • Pacific Coast Climate Collaborative; http://pacificcoastcollaborative.org • South Coast Climate Science Alliance; http://www.climatesciencealliance.org • NOAA RISA teams (CLIMAS, CNAP, Western Water Assessment); http://www.climas.arizona.edu; https://scripps.ucsd.edu/programs/cnap/; http://wwa.colorado.edu • Southwest Climate Science Center; http://www.swcsc.arizona.edu • California Ocean Science Trust; http://www.oceansciencetrust.org • Pacific Institute; https://pacinst.org
SOURCES OF CLIMATE DATA	• CalAdapt (with an emphasis on climate change data for users); http://cal-adapt.org • Western Regional Climate Center; https://wrcc.dri.edu • California LCC "Climate Commons"; http://climate.calcommons.org • Southern California Coastal Observing System (SCOOS); http://www.sccoos.org • California Cooperative Oceanic Fisheries Investigations (CalCOFI); http://calcofi.org

26: ALASKA

#	Page/Line	Comment
437	P1170/L33-34	There is clumsy wording of the text "qualitative … ecosystems" and is difficult to understand. This differs from many of the relatively simple definitions in the literature.

#	Page/Line	Comment
438	P1170/L37-39	The range of adaptations that are underway, as stated in this sentence, are not described in the adaptation section of this chapter.
439	P1171/L2-6	It is unclear how Figure 26.4 relates to the text that precedes or follows it.
440	P1172/L1-33	Add citations to the introduction.
441	P1174/L8	The text says "(c) same as (c)," but it should say "(d) same as (c)."
442	P1174/L29	Do "coastal areas" refer to all coastal areas of the state, including southeastern Alaska?
443	P1177/L8	Is it correct that fish feed on pteropods rather than pteropods feeding on fish?
444	P1179/L11	The cited Mann et al. 2012 did not discuss shrubs. They talk about the conversion of conifer forests to deciduous vegetation (including shrublands and deciduous forests). They emphasize the importance of aspen forests (not shrublands) as a future state that will become more common. See also comment #447.
445	P1179/L18	What about the effects of permafrost thaw on the water-quality impacts of large mines?
446	P1184/L24	Change "quality" to "air quality."
447	P1186/L30-32	The same sentence is repeated twice.
448	P1196/L2	Is "Arctic ice sheet," meant to be "sea ice"? Clarification is needed.
449	P1196/L34	The conclusion by Mann et al. 2012 referred to changes from conifer to deciduous cover, not changes from forest cover to shrubs, as stated in this sentence (and in an earlier reference to the paper by Mann et al.). The statement that forests are changing to shrublands may be true, but the Mann et al. 2012 reference does not support this statement. See comment #442.
450	P1197/L6-8	The sentence "Thermal … regions" seems to have words missing and is unclear.
451	P1201/L1-2	There is a word missing from this sentence.

Appendix 5: FREQUENTLY ASKED QUESTIONS

#	Page/Line	Comment
452	P1444/L35-36	A better topic sentence would be "Additional lines of evidence support the idea that the world is warming."

#	Page/Line	Comment
453	P1445/L7-9	Because the FAQ focus is on *how we know* the Earth is warming and not *why* the Earth is warming, reference to human activities being responsible for the warming does not seem appropriate.
454	P1444/L16	The link included on page 1444, line 16 should be more clearly noted as an external source. As currently written, it gives the impression that it links to Figures A5.1 and A5.2 in the draft NCA4.
455	P1451/L11	The text needs a comma after "heat."
456	P1451/L13	Define the abbreviation GHG before using it (if this line is kept in the text).
457	P1453/L2	The word lead "lead" should be changed to "leads" or "must lead."
458	P1447/L17	Insert the word "rapidly" so that it reads "use of coal, oil, and gas has rapidly changed the atmosphere."
459	P1451/L15	The word "similar" should be changed to "similarly," because the physical mechanism is actually different (greenhouse versus greenhouse gases). Alternatively, the comparison could be omitted.
460	P1459/L13	It would be appropriate to say "very strong regional effects in some areas."
461	P1459/L13-16	This text is likely too technical for the intended audience and should be revised.
462	P1470/L14	The text currently says "scientists compare data." which could be revised to say "scientists evaluate data."
463	P1471/L21	Edit text to say "at a given location over periods of multiple years to decades."
464	P1473/L21	The word "arctic" should be capitalized.
465	P1475/L36	Insert "noted above" so that the text says "scientific evidence noted above indicates that."
466	P1476/L5	Insert "human-caused" so that the text says "of human-caused global warming on these events."
467	P1476/L12	It is suggested that the words "the severity of" and "some" be added, so that the text reads "global warming has contributed to the severity of some individual weather and climate events."
468	P1476/L15	The text that says "we can model" is jargon. It is suggested that "model" be replaced with "simulate."
469	P1477/L3	Insert "human-caused" so that the text says "to detect the influence of human-caused global warming."
470	P1477/L4	Insert "events" so that the text states "and, to a lesser extent, heavy rainfall events, is better at present."

#	Page/Line	Comment
471	P1477/L7-8	It is suggested that the text be changed to the following wording: "ability to attribute how much human-caused global warming contributes to specific weather and climate events."
472	P1477/L12	Insert "human-caused" so that the text says "link human-caused global warming to particular weather and."
473	P1477/L15	Insert "human-caused" so that the text says "while human-caused global warming contributed."
474	P1477/L17	Change the wording to: "activity, but human-caused global warming leads to."
475	P1477/L33	Clarification is needed here. A change in wording to: "exist, only that the data record is not long enough." is suggested.
476	P1488	The intermediate scenario referenced in Figure A5.29 should be consistently labeled throughout the report and FAQs.
477	P1488/L16	The text should say "Sea level is expected to continue rising at an accelerating rate this century, increasing the frequency of nuisance flooding, *as* well as intensifying coastal."

Appendix 5-Suggested References

FAQ: WHY ARE SCIENTISTS SO CERTAIN THAT HUMAN ACTIVITIES ARE THE PRIMARY CAUSE OF RECENT GLOBAL WARMING?	• Marcott, S. A., J. D. Shakun, P. U. Clark, and A. C. Mix (2013), A reconstruction of regional and global temperature for the past 11,300 years, *Science, 339,* 1198-1201. • Marsicek, J., B. N. Shuman, P. J. Bartlein, S. L. Shafer, and S. Brewer (2018), Reconciling divergent trends and millennial variations in Holocene temperatures, *Nature, 554*, 92.
FAQ: WHAT ROLE DOES WATER VAPOR PLAY IN GLOBAL WARMING?	• Myhre, G., D. Shindell, F.-M. Bréon, W. Collins, J. Fuglestvedt, J. Huang, D. Koch, J.-F. Lamarque, D. Lee, B. Mendoza, T. Nakajima, A. Robock, G. Stephens, T. Takemura, and H. Zhang (2013), Anthropogenic and Natural Radiative Forcing, in *Climate Change 2013: The Physical Science Basis. Contribution of Working Group I to the Fifth Assessment Report of the Intergovernmental Panel on Climate Change,* T. F. Stocker, D. Qin, G.-K. Plattner, M. Tignor, S. K. Allen, J. Boschung, A. Nauels, Y. Xia, V. Bex, and P. M. Midgley, eds., Cambridge University Press, Cambridge, United Kingdom and New York, NY, USA.

FAQ: WHAT IS THE SOCIAL COST OF CARBON?	• NASEM (2017), *Valuing Climate Damages: Updating Estimation of the Social Cost of Carbon Dioxide,* Washington, DC, The National Academies Press.
FAQ: HOW FAST ARE GLACIERS MELTING IN GLACIER NATIONAL PARK?	• Pederson, G. T., D. B. Fagre, S. T. Gray, and L. J. Graumlich (2004), Decadal-scale climate drivers for glacial dynamics in Glacier National Park, Montana, USA, *Geophysical Research Letters, 31,* L12203, doi:10.1029/2004GL019770. • Pederson, G. T., L. J. Graumlich, D. B. Fagre, T. Kipfer, and C. C. Muhlfeld (2010), A century of climate and ecosystem change in Western Montana: What do temperature trends portend?, *Climatic Change, 98,* 133-154. • Pederson, G. T., S. T. Gray, C. A. Woodhouse, J. L. Betancourt, D. B. Fagre, J. S. Littell, E. Watson, B. H. Luckman, and L. J. Graumlich (2012), The unusual nature of recent snowpack declines in the North American cordillera, *Science,* 333(6040), 332-335. • Pederson, G. T., S. T. Gray, T. Ault, W. Marsh, D. B. Fagre, A. G. Bunn, C. A. Woodhouse, and L. J. Graumlich (2011), Climatic controls on the snowmelt hydrology of the northern Rocky Mountains, *Journal of Climate, 24,* 1666-1687. • Pederson, G. T., J. L. Betancourt, and G. J. McCabe (2013), Regional patterns and proximal causes of the recent snowpack decline in the Rocky Mountains, US, *Geophysical Research Letters, 40,* doi: 10.1002/grl.50424.
FAQ: IS CLIMATE CHANGE AFFECTING U.S. WILDFIRES?	• Abatzoglou, J. T., and A. P. Williams (2016), Impact of anthropogenic climate change on wildfire across western US forests, *Proceedings of the National Academy of Sciences, 113*(42),11770-11775. • Jolly, W. M., M. A. Cochrane, P. H. Freeborn, Z. A. Holden, T. J. Brown, G. J. Williamson, and D. M. J. S. Bowman (2015), Climate-induced variations in global wildfire danger from 1979 to 2013, *Nature Communications, 6,* 7537. • Schoennagel, T., J. K. Balch, H. Brenkert-Smith, P. E. Dennison, B. J. Harvey, M. A. Krawchuk, N. Mietkiewicz, P. Morgan, M. A. Moritz, R. Rasker, M. G. Turner, and C. Whitlock (2017), Adapt to more wildfire in western North American forests as climate changes, *Proceedings of the National Academy of Sciences,* doi:10.1073/pnas.1617464114. • Westerling, A. L. (2016), Increasing western US forest wildfire activity: Sensitivity to changes in the timing of spring, *Philosphical Transactions of the Royal Society B: Biological Sciences, 371*(1696), 20150178.

Appendix C. Statement of Task

An *ad hoc* committee will conduct a review of the draft Fourth National Climate Assessment (NCA4), concurrent with the public comment period. The review will address the following questions about the draft report:

1. Does the report meet the requirements of Section 106 of the Global Change Research Act?
2. Do the key messages reflect current understanding about observed and projected impacts to the United States, the challenges, opportunities and success stories for addressing risk, and identification of emerging issues related to climate change?
3. Does the report accurately reflect the peer-reviewed scientific literature, with a particular focus on literature since the last National Climate Assessment (i.e., since ~2013)? Are there any critical content areas missing from the report?
4. Are the findings documented in a consistent, transparent and credible way?
5. Is the report written at a technical level that is appropriate for the intended audience?
6. Are the report's key messages and graphics clear, internally consistent, and appropriate? Specifically, do they reflect supporting evidence, include an assessment of likelihood, and communicate effectively?
7. Are the data and analyses handled in a consistent, transparent, and credible manner? Are statistical methods applied appropriately?
8. What other significant improvements, if any, might be made in the document?

Appendix D. Committee Biographies

DR. ROBIN E. BELL (*Chair*) is Palisades Geophysical Institute/Lamont Research Professor at Columbia University's Lamont-Doherty Earth Observatory. She is also currently President-elect of the American Geophysical Union. For 35 years, Dr. Bell has worked alongside a team of Lamont-Doherty Earth Observatory scientists and engineers to coordinate nine major aerogeophysical expeditions to Antarctica and Greenland in order to study ice sheet collapse. Dr. Bell was a leading proponent of the 2007-2008 International Polar Year and has chaired the National Academies' Polar Research Board. Her work examines the implications of climate change on the poles and involves adapting scientific instruments to produce imaginative new insights into the Polar regions. She also conducts work focused on estuarine processes and led a Lamont team to map the Hudson River from Staten Island to Albany. Dr. Bell received her Ph.D. in geophysics from Columbia University in 1989. She has been part of the research staff at Columbia University's Lamont-Doherty Earth Observatory since 1989 and is a member of the Earth Institute faculty. Dr. Bell has published more than 90 peer-reviewed articles and more than 30 other publications, and continues to pursue new directions in her field to meet the challenges presented by climate change in the Polar regions.

DR. BILAL M. AYYUB is Professor and Director of the Center for Technology & Systems Management in the Department of Civil and Environmental Engineering at the University of Maryland. Dr. Ayyub's main research interests and work are in risk, uncertainty and decision analysis, resilience, sustainability and adaptation to a changing climate, and systems engineering applied to civil, infrastructure, energy, defense and maritime infrastructure. His work includes risk-informed planning and decision making covering several aspects of project lifecycles and portfolios, such as sustainment and life expectancy assessment of systems, protection of critical infrastructure, bidding strategies, project execution risk, operational risk, risk management including risk transfer and risk finance, and liability and exposure analysis. Professor Ayyub is a distinguished member of the American Society of Civil Engineers (ASCE). He is also a fellow of the American Society of Mechanical Engineers (ASME), the Society of Naval Architects and Marine Engineers (SNAME) and the Society for Risk Analysis (2017-18 Treasurer), and a senior member of the Institute of Electrical and Electronics Engineers (IEEE). He is the FY2017-18 Chair of the ASCE Infrastructure Resilience Division (IRD), and the FY2017 Chair of the ASCE Committee on Adaptation to a Changing Climate (CACC). He chaired the ASME Safety Engineering and Risk Division (SERAD). He is currently the Editor-in-Chief of the ASCE-ASME J. of Risk and Uncertainty in Engineering Systems, and on the editorial boards of several journals, and chaired several ASCE and non-ASCE conferences and workshops. He is the author of over 300 refereed papers and over 30 books and edited volumes. He earned a doctorate degree from the Georgia Institute of Technology (1983).

DR. MICHELLE L. BELL is a professor of environmental health at the Yale University School of Forestry and Environmental Studies, with secondary appointments at the Yale School of Public Health, Environmental Health Sciences Division and the Yale School of Engineering and Applied Science, Environmental Engineering Program. Dr. Bell's research investigates how human health is affected by atmospheric systems, including air pollution and weather. Much of this work is based in epidemiology, biostatistics, and environmental engineering. The research is designed to be policy-relevant and contribute to well-informed decision-making to better protect human health. She is the recipient of the Prince Albert II de Monaco/Institut Pasteur Award, Rosenblith New Investigator Award, and the NIH Outstanding New Environmental Scientist (ONES) Award. Dr. Bell received her Ph.D. in environmental engineering from John Hopkins University in 2002.

DR. DANIEL G. BROWN is a professor and director in the School of Environmental and Forest Sciences at the University of Washington. His work, published in over 200 refereed articles, chapters, and proceedings papers, has aimed at understanding human-environment interactions through a focus on land-use and land-cover changes, through modeling these changes, and through spatial analysis and remote sensing methods for characterizing landscape patterns. Recent work has used agent-based and other spatial simulation models to understand and forecast landscape changes that have impacts on carbon storage and other ecosystem services, and human health and well-being. He has conducted field work on three different continents: Africa, Asia, and North America. He has chaired the Land Use Steering Group and Carbon Cycle Steering Group and was a lead coordinating author for the third National Climate Assessment, all under the auspices of the U.S. Climate Change Science Program. He has served on the following National Academies committees: "Needs and Research Requirements for Land-Change Modeling" (chair), "Mapping Sciences Committee" (member), "Earth Science and Applications from Space" (member). In addition, he has served as a member of the NASA Land Cover and Land Use Change Science Team, as panelist for NASA, EPA, USDA Forest Service, the National Science Foundation, and the European Research Council, and on the Editorial Boards for Landscape Ecology, Computers, Environment and Urban Systems; International Journal of Geographical Information Science; and the Journal of Land Use Science. In 2009 he was elected fellow of the American Association for the Advancement of Science. Dr. Brown earned his Ph.D. in geography in 1992 from the University of North Carolina at Chapel Hill.

DR. DANIEL R. CAYAN is a Research Meteorologist at the Scripps Institution of Oceanography, University of California San Diego. His work is directed at understanding climate variability and changes over the Pacific Ocean and North America and climate impacts on water, wildfire, health, and agriculture in California and western North America. Dr. Cayan is co-lead for the California Nevada Applications Program, which delivers climate information to decision makers in California and Nevada. He is also one of the scientists involved in the Southwest Climate Science Center, sponsored by the Department of Interior through the U.S. Geological Survey. Cayan is a fellow of the American Geophysical Union. He received a B.S. in meteorology and oceanography from the University of Michigan and a Ph.D. in oceanography from the University of California, San Diego.

DR. F. STUART CHAPIN III (NAS) is a professor emeritus of the University of Alaska Fairbanks. Dr. Chapin's research addresses the effects of changes in climate and wildfire on Alaskan ecology and rural communities. He explores ways that communities and agencies can develop options that increase sustainability of ecosystems and human communities over the long term despite rapid climatic and social changes. Through projections of future climate, ecology, and subsistence resources, his research helps people make more informed choices about options for long-term sustainability. Also, his research in earth stewardship explores ways that society can proactively shape changes toward a more sustainable future through actions that enhance ecosystem resilience and human well-being. Dr. Chapin pursues this internationally through the Resilience Alliance, nationally through the Ecological Society of America, and in Alaska through a community partnership that links the sustainability visions of rural indigenous communities with university research expertise to implement those visions. Dr. Chapin received his B.S. in biology from Swarthmore College in 1966 and his Ph.D. in biology from Stanford University in 1973. Dr. Chapin was elected as a member of NAS in 2004. He has participated in several activities with the National Academies including, most recently, serving on the Board on Environmental Change and Society (2012-present) and participating as a member of the Committee on Preparing for Nine Billion on the Planet (2013) and the Committee to Review of U.S. Landscape Conservation Cooperatives (2015-16).

DR. JOHN B. GATES is currently a senior scientist at The Climate Corporation where he serves as Lead for Soil and Crop Modeling. His work leverages data science, numerical modeling and cloud computing to address precision agricultural challenges including nutrient management and risks from extreme weather. Dr. Gates' expertise lies in digital agriculture, nitrogen and irrigation management, soil fertility and hydrology. Dr. Gates received his B.S. in mathematical sciences from the University of Arkansas at Fayetteville in 2002, his M.Sc. in environmental change and management in 2003, and his Ph.D. in geography and the environment in 2007, both from the University of Oxford. He was a postdoctoral fellow at the University of Texas at Austin until 2009 and then served as Harold and Esther Edgerton Assistant Professor in the Department of Earth and Atmospheric Sciences at the University of Nebraska-Lincoln until 2014. He was an associate editor for Hydrogeology Journal from 2009-2014, and has been a Global Fellow of the Robert B. Daugherty Water for Food Institute since 2010.

DR. L. RUBY LEUNG (NAE) is a Battelle Fellow at Pacific Northwest National Laboratory and an Affiliate Scientist at National Center for Atmospheric Research. Her research broadly cuts across multiple areas in modeling and analysis of climate and water cycle including orographic processes, monsoon climate, climate extremes, land surface processes, land-atmosphere interactions, aerosol-cloud interactions, and human-Earth system interactions. She is the Chief Scientist of the U.S. Department of Energy (DOE)'s Energy Exascale Earth System Model (E3SM). She has been actively involved in development and application of regional climate models as well as evaluation and analysis of high resolution and variable resolution global climate models. She served on the National Academies committee on "A National Strategy for Advancing Climate Modeling" in 2012 and provided reviews on several National Academies' reports. Currently she is a member of National Academies Board on Atmospheric Sciences and Climate (BASC), DOE Biological and Environmental Research Advisory Committee (BERAC), National Oceanic and Atmospheric Administration (NOAA) Climate Working Group (CWG) of the Science Advisory Board (SAB), and the advisory panel for the National Center for Atmospheric Research (NCAR) Mesoscale and Microscale Meteorology (MMM) Laboratory. She has organized many workshops to define research needs and directions in areas such as integrated water cycle, regional climate modeling, and mountain hydroclimate and water resources. Dr. Leung is a member of the National Academy of Engineering and Washington State Academy of Sciences. She is also a fellow of the American Association for the Advancement of Science (AAAS), American Geophysical Union (AGU), and American Meteorological Society (AMS). She received a BS in Physics and Statistics from Chinese

University of Hong Kong and an MS and PhD in Atmospheric Sciences from Texas A&M University.

DR. JANET PEACE is the Senior Vice President of Policy and Business Strategy at the Center for Climate and Energy Solutions (C2ES). As part of a three-person executive management, she oversees the center's domestic climate policy programs, its Business Environmental Leadership Council, its climate science and resilience program and analysis of market-based policy options. Dr. Peace brings more than 25 years and a wide spectrum of experience on environmental issues to her work at C2ES. As a recognized expert on climate policy, she is a member of the Program Advisory Board for American University's Center for Environmental Policy, a member of the External Advisory Committee for ASU's Urban Resilience to Extremes Sustainability Research Network and a past member of both the National Academies' Roundtable on Climate Change Education and the Council of Canadian Academies on oil sands environmental technologies. Prior to C2ES, Dr. Peace worked on climate policy in Alberta and taught environmental and natural resource economics at the University of Calgary. She also worked as a resource specialist with the U.S. General Accounting Office and as a geologist with the U.S. Geological Survey. She holds a Ph.D. (1994) and Master of Science (1992) in Mineral Economics from Colorado School of Mines and an undergraduate degree in geology (1985) from University of Colorado.

DR. JULIE PULLEN is an associate professor in civil, environmental, and ocean engineering at Stevens Institute of Technology. She holds a joint appointment with Brookhaven National Laboratory and is an adjunct research scientist at Columbia's Lamont-Doherty Earth Observatory. Previously, she was the Director of the Department of Homeland Security (DHS) National Center of Excellence in Maritime Security at Stevens and a former science fellow at Stanford's Center for International Security and Cooperation. She studies complex coastal air/sea interactions utilizing high-resolution (<5 km) coupled ocean/atmosphere/hydrology models and observations from targeted field campaigns around the globe. Dr. Pullen has served on the steering team for field studies in urban air contaminant dispersion (DHS/Defense Threat Reduction Agency NYC Urban Dispersion Program) and tropical meteorology and oceanography (Office of Naval Research, PhilEx and PISTON programs). Dr. Pullen's research contributes to the understanding and development of resilience and sustainability in coastal environments, and the enhancement of Earth System Models on weather, subseasonal-to-seasonal, and climate timescales. Her work also improves the treatment of air/sea/urban processes within transport and dispersion models for chemical/biological radiological and nuclear (CBRN) applications. In 2015 Dr. Pullen was elected as the physical oceanography councilor for The Oceanography Society. She was a member of the 2014-2016 National Academies committee on Subseasonal to Seasonal Earth System Prediction and is on the international GODAE Coastal Ocean and Shelf Seas Task Team. Dr. Pullen is a board member of the Waterfront Alliance, a civic organization representing more than 1,000 groups with a stake in the NY/NJ waterfront, and is co-chair of the policy committee. She serves on the Science Advisory Committee for the Environmental & Climate Sciences Department of Brookhaven National Laboratory. Dr. Pullen holds a master's degree in applied mathematics from the University of Arizona, and a Ph.D. in physical oceanography from Oregon State University.

DR. LaVERNE E. RAGSTER is a retired professor of marine biology and President Emerita of the University of the Virgin Islands (UVI). She has conducted research and training in the areas of algal physiology and natural resource management, presented and published in the areas of plant physiology, natural resource management and training, and obtained training and practical experience in institutional and leadership development over a 35-year span. She has served on national, Caribbean, and local boards and commissions addressing higher education, environmental justice, waste management, natural resource management, fisheries, ocean observing systems, and sustainable development. Her current scholarly work at the UVI Caribbean Exploratory (NIMHD) Research Center addresses climate change adaptation and linkages to public health in the Caribbean. Dr. Ragster has been recognized and honored by Virgin Islands organizations, national organizations, higher education institutions, and the President of the United States for her teaching, work in conservation, leadership and community service. Her educational career included completion of a B.S. in biology and chemistry (University of Miami) in 1973, a M.S. in biology (San Diego State University-algal physiology concentration) in 1975 and a Ph.D. in biology (University of California, San Diego-plant biochemistry concentration) in 1980. During the last eight years she has published a number of papers on the role of natural resources in resource management and development, produced programs for the training of faculty and resource managers, and developed curriculum materials to teach natural resource management at the university level in the Caribbean.

DR. MARY RUCKELSHAUS is the Director of The Natural Capital Project and a consulting professor at Stanford University. Dr. Ruckelshaus previously led the Ecosystem Science Program at NOAA's Northwest Fisheries Science Center in Seattle, WA. Prior to that, she was an Assistant Professor of biological sciences at The Florida State University. The main focus of her recent work is on developing a universal approach for valuing nature and mainstreaming it into high-leverage decisions globally. Dr. Ruckelshaus serves on the Science Council of The Nature Conservancy and is a Trustee on its Washington Board, is a member of the U.S. Ocean Research Advisory Panel—charged with providing independent science advice to the National Ocean Council—and is a past chair of the Science Advisory Board of the National Center for Ecological Analysis and Synthesis (NCEAS). She was Chief Scientist for the Puget Sound Partnership, a public-private institution charged with achieving recovery of the Puget Sound terrestrial, freshwater and marine ecosystems. Dr. Ruckelshaus has a bachelor's degree in human biology from Stanford University, a master's degree in fisheries from the University of Washington, and a doctoral degree in botany, also from Washington.

MS. SUSANNE TORRIENTE is chief resilience officer for the city of Miami Beach. She is a seasoned public administrator with more than 26 years of service in Miami Dade County and several signature cities in the region. She joined the City of Miami Beach in September 2015 as Assistant City Manager (ACM) and Chief Resiliency Officer (CRO). Her sustainability and resiliency portfolio includes planning, building, code compliance, and environment and sustainability. The City of Miami Beach is a leader and pioneer in adaptation efforts in South Florida. As CRO she is leading the effort to develop an action-oriented citywide resiliency strategy and a broader Greater Miami & the Beaches 100 Resilient Cities Strategy supported by the Rockefeller Foundation. She is also on the board of the Association of Climate Change Officers (ACCO). During her four years as assistant city manager for operations in Fort Lauderdale, Ms. Torriente successfully focused on streamlining city services, initiating strategic management practices & performance measurement systems, and integrating sustainability into city operations. She was the lead staff to the city's visioning process producing Fast Forward

Fort Lauderdale: Our City, Our Vision, 2035. She was also selected as a Fellow for the 2012 ICMA Sustainable Communities Fellowship. In 2009, Ms. Torriente was appointed Miami-Dade County's first Sustainability Director and became a founding member of the Southeast Florida Regional Climate Compact staff steering committee. Ms. Torriente earned her Master's in public administration from the University of Miami in 1990.

DR. ELKE U. WEBER is the Gerhard R. Andlinger Professor in Energy and the Environment and a professor of psychology and public affairs at Princeton University. Dr. Weber's research interests include determinants of risky decision making, cultural, age, gender, and species differences in decision making under risk and uncertainty, measurement of risk perception and risk attitude, role of memory and other cognitive processes in preference construction, financial decision making, environmental decision making, decision neuroscience, and neuroeconomics. Dr. Weber is a fellow of the American Academy of Arts and Sciences, the American Psychological Association, the American Psychological Society, the Society for Experimental Psychology, and the Society for Risk Analysis. She also served previously as president of the Society for Judgment and Decision Making, the Society for Neuroeconomics, as well as the Society for Mathematical Psychology and is a member of the German National Academy of Sciences. Dr. Weber earned her B.A. in psychology from York University in 1980, and her Ph.D. in behavior and decision analysis from Harvard University in 1984.

DR. CATHY WHITLOCK is a professor of Earth Sciences at Montana State University and a fellow of the Montana Institute on Ecosystems. Dr. Whitlock's research interests include Quaternary environmental change, paleoecology and paleoclimatology with a focus on vegetation, fire, and climate history. She is nationally and internationally recognized for her scholarly contributions and leadership activities in the field of past climatic and environmental change, and she has published over 190 reviewed journal articles and book chapters on this topic. Her current research sites extend from Yellowstone and the western U.S. to New Zealand, Tasmania, and Patagonia. Since her arrival at MSU in 2004, Dr. Whitlock has built a successful research and teaching program, and the MSU Paleoecology Lab supports post-docs, graduate students, and undergraduates and visiting scientists from around the world. Her research has been funded by grants from the National Science Foundation, Joint Fire Sciences Program, National Park Service, Department of Energy, USDA Forest Service, and US Geological Survey. She is past President of the American Quaternary Association and has served on national and international advisory committees concerned with climate change. Dr. Whitlock is also the lead author of the 2017 Montana Climate Assessment. She earned her Ph.D. in geological sciences from the University of Washington in 1983 and her BA from Colorado College in 1975.

DR. GARY YOHE is the Huffington Foundation Professor of Economics and Environmental Studies at Wesleyan University; he has been on the faculty at Wesleyan for more than 30 years. He was educated at the University of Pennsylvania (in mathematics – 1970), and received his PhD in economics from Yale University in 1975. He is the author of more than 150 scholarly articles, several books, and many contributions to print and television coverage of climate issues. He began his work on climate change in 1982. Most of his current work has focused attention on the mitigation and adaptation/impacts sides of the climate issue. Involved since the early 1990's with the Intergovernmental Panel on Climate Change, he received a share of the 2007 Nobel Peace Prize and has been involved in their work supporting global negotiations under the United

Nations Framework Convention on Climate Change. Professor Yohe serves as a member of the New York City Panel on Climate Change. He has testified before the Senate Foreign Relations Committee on the "Hidden (climate change) Cost of Oil," the Senate Energy Committee on the Stern Review, and the Senate Banking Committee on "Material Risk from Climate Change and Climate Policy." He served as a member of the Adaptation Panel of the National Academies initiative on America's Climate Choices and as a member of a National Academies Committee on Stabilization Targets for Atmospheric Greenhouse Gas Concentrations that was chaired by Susan Solomon (among many other appointments). Professor Yohe was also Vice Chair of the National Climate Assessment Development and Advisory Committee for Third National Climate Assessment for the Obama Administration that was released by the White House in May of 2014. He currently serves in his seventh year as co-editor of Climatic Change with Michael Oppenheimer.